MATHEMATICS
FOR
EVERYMAN

FROM SIMPLE NUMBERS TO THE CALCULUS

Egmont Colerus

Translated by
B. C. AND H. F. BROOKES

DOVER PUBLICATIONS, INC.
Mineola, New York

Bibliographical Note

This Dover edition, first published in 2002, is an unabridged republication, with a short list of errata, of the work published by Emerson Books, Inc., New York, in 1958. This edition is a translation of the original work, published in Germany under the title *Vom Einmaleins zum Integral: Mathematik für Jedermann* by P. Zsolnay, Berlin, in 1937.

Library of Congress Cataloging-in-Publication Data

Colerus, Egmont 1888-1939
    [Vom Einmaleins zum Integral. English]
    Mathematics for everyman : from simple numbers to the calculus / Egmont Colerus ; translated by B. C. and H. F. Brookes.
        p. cm.
    Originally published: New York : Emerson Books, 1957.
    Includes bibliographical references and index.
    ISBN 0-486-42545-2 (pbk.)
       1. Mathematics. I. Title.

QA37.3 .C6513 2002
510—dc21

2002031226

Manufactured in the United States of America
Dover Publications, Inc., 31 East 2nd Street, Mineola, N.Y. 11501

CONTENTS

TRANSLATORS' NOTE

WE have tried to convey the spirit and enthusiasm of the original German rather than to make an exact or literal translation. At the same time we were asked by the present publishers to reduce the length of the book by one third. We believe that in making this reduction we have also been able to make the book more attractive to English readers ; the reduction was achieved mainly by simplifying the illustrative examples and by shortening some of the mathematical arguments.

B. C. B.
H. F. B.

NOTE ON THE AUTHOR

IN German-speaking countries, Egmont Colerus has had great success with his popular works on mathematics. He is able to appreciate " everyman's " difficulties in understanding mathematics because, as a young man, he taught himself the subject and clearly remembered his own difficulties. He undertakes his task with great thoroughness but with an uncommon lightness of touch, and an unconventional approach. Though his work is addressed to " everyman " who would like a better understanding of mathematics, those who wish to approach this fascinating subject more seriously will find that the coherence and order of their mathematical knowledge is greatly strengthened by reading this book.

There are particularly illuminating treatments of Number and of Mathematical Operations. The author's interest in mathematical history and philosophy gives a breadth and depth to this book which is unusual in works of such a popular nature.

AUTHOR'S PREFACE

MATHEMATICS is a trap. If you are once caught in this trap you hardly ever get out again to find your way back to the original state of mind in which you were before you began to investigate mathematics. It would take far too long to expound the reasons for this. I shall content myself with establishing the results which follow from it.

The first result of the " trap-like " quality of mathematics is that there is a great shortage of mathematics teachers. The ability to grasp the subject yourself and to explain it clearly to others is not often given to any one person. This gives rise to the mathematical inferiority complex which is so frequently found in educated people and in those who are well disposed towards education. Please do not misunderstand me. I am not attacking anybody ; on the contrary, I am trying to defend myself, for a layman attempts to expound this most difficult of all branches of knowledge only on very rare occasions.

For years I have noted my own difficulties in mathematics and those of my colleagues too. I conceived the plan to write down at a fairly elementary level my experiences whilst learning the subject. For, according to my own analysis of the " trap," it is to be feared that in a few years' time I too will be unable to find my way out of it again.

There is another very cogent reason for writing down these experiences. It is obvious that mathematics, its methods and concepts are becoming an increasingly important part of all branches of science, one might almost say of everyday life. It is an extremely unsatisfactory state of affairs, almost amounting to a scandal, that a reader should be frightened and put off by a row of hieroglyphics in the middle of a serious treatise or that he should have to let a small number of the initiated finish their reading whilst he can only stand by and shrug his shoulders. I am not talking about anything on such a high level as the theory of relativity or the quantum theory, but of mathematics which might appear in any medical or economics publication ; I am leaving entirely out of account statistics which has become a completely mathematical subject, especially in Anglo-Saxon countries. Besides, mathematics crops up much more slyly in everyday speech. We read in the newspapers about "average temperatures," "mean values," "optimum . . ." "fields of force," and similar expressions borrowed from mathematics and mathematical physics.

It is not necessary to hear these words without understanding them, and, as a result, to appear to be an uneducated or inferior being. If we are willing to take pains these words will cease to sound imposing and will become comprehensible. The one pre-requisite is an inflexible determination to learn. When in Alexandria in about the year 300 B.C. the greatest of the Greek geometers, Euclid, was asked by his king Ptolemy Philadelphus for an easy way to learn mathematics, he bravely replied, "There is no royal road to mathematics." Anyone who has just a superficial knowledge of the subject but who knows its purely intellectual character, built up as it is step by step, will agree with Euclid. But this need not lead to defeatism and despair. There are endless possibilities lying between the easy path and the stony one.

Most books on mathematics, admirable as they may be, are often unintelligible to the ordinary man because they assume in their reader a secondary school education and a knowledge of the elementary mathematical processes. I myself suffered from this assumption on the part of my teachers when I decided to continue my education and to study higher mathematics and statistics. In this book I am attempting to remove such difficulties so that no one need suffer from an inferiority complex as regards mathematics, and I shall deem it a triumph if my readers continue their studies and end by despising my wordy explanations and my simplicity. If any of my readers use this book as a "coach," and should any contradictions arise between what the book says and what your teacher says, your teacher is always right!

It is my pleasant duty to thank my own teacher, Dr. Walther Neugebauer, for having made me for the first time aware of the greatness of mathematics and for having shown me the core of the whole matter. If I succeed in conveying to my readers something of this greatness, something of that quality in mathematics which made a poet once write that pure mathematics is religion, then I shall be very happy. Unfortunately, a feeling of inferiority about mathematics sometimes arouses resentment and hatred. The great Greek woman mathematician, Hypatia, the only woman in the front rank of mathematicians, was stoned by the people—not on account of religious fanaticism but because of a hatred of something they did not understand. This book is an attempt to combat the general aversion for mathematics.

EGMONT COLERUS.

CHAPTER I

NUMBERS

A PATIENT is sitting in the doctor's waiting-room. He suspects that he won't get away very quickly so he decides to while away the time by reading. There are all kinds of brochures and advertisements of resorts and cruises on the table. He is particularly attracted by a picture which reveals all the wonder of southern seas and tropical townships. His interest is aroused and he opens the booklet—but he is very disappointed. He can hardly understand a word ; it is written in Portuguese. He can understand what the pictures are about and there is something else too that he can follow without a translation, that is the columns of figures, the tables of statistics and the departure and arrival times.

You will think that this is rather a childish example of something which is quite self-evident. Who has ever doubted that nowadays almost all civilised countries use a common system of numbers ? What is there so remarkable or worrying about this booklet ? The number 3 in a Portuguese book means exactly the same as 3 in an English one, and so does a sum like $521 \times 7 = 3,647$. You would think that there is nothing more to be said about it. On the contrary ; if we look more closely at the rather trivial example we shall come at once on the deepest puzzles of mathematics and we shall grasp a number of most important fundamental principles.

There is one thing that has so far been overlooked. When a Portuguese or a German reads the numbers in the prospectus, each uses different words for them. The way in which the numbers are pronounced in various languages goes to the very root of the mysteries of the number system. The English say " twenty-four," the Germans " four and twenty." Instead of saying " septante " for 70, which would be the logical sequence of " quarante, cinquante, soixante . . ." the Frenchman says rather surprisingly " soixantedix " and for 80 he actually undertakes some multiplication in " quatrevingt," or 4×20.

The first thing we must note is that there is only a very slight connection between our international system of numbers and the alphabet system. The two are based on quite different principles. Numbers are each in themselves symbols standing for an idea ; the letters of the alphabet do not in themselves represent ideas but only sounds ; only when they are strung together as words can they become symbols for ideas. The concept " 3 " needs one sign when written as a number. When written as the word " three," 5 letters are used.

All this is only the beginning. We have been talking up till now of figures and numbers, not of the great number system which is said to be the greatest pride of the human intellect. At this point some readers will say, " We know all about the number system if what is meant by that is simply a system scaled in tens. We use it ourselves every day and if this book is going to theorise instead of sticking to necessary explanations we will just shut it up and fling it away." In defence it must be said that no attempt will be made to put forward a theory of numbers or to explain why three hundred is written " 300." Because nothing should be taken for granted however, we must seize hold of such matters which are generally known and accepted, in order to be able to make some of the higher concepts of mathematics intelligible right from the start.

The reader would do wrong to speak slightingly of our system of numbers scaled in tens. One of its greatest merits is that it can be learnt by a Primary School child. But there was a time in history when what a child can now do at school was a difficult problem for the greatest mathematicians. For the well-nigh automatic working of our present number system had not yet been discovered or developed. The technique of calculation only became commonly known in the West during the eleventh century A.D. At that time two " schools of reckoning " were struggling for precedence ; the one used the abacus, the other the devices of Muhammad ibn Ahmad al-Khwarizmi. The abacus is an ancient counting frame. Imagine that we have before us a board divided up by vertical lines. Each column represents a step in the system, a single number, a ten, a hundred, a thousand, etc. To use the abacus to reckon with, we place the required number of counters in each column. Suppose we had to add 504,723 and 609,802. We place the first number on the board in the correct columns, using white counters. Then we do the same with the black counters for the second number. We have to count up the total of black and white counters to arrive at the result.

There is no zero (0) in this abacus system. When adding up the counters we must not forget that 1,500 equals 1 thousand, 5 hundred; 104,000 equals 1 hundred-thousand and 4 thousand. It will soon be obvious that al-Khwarizmi's system of numbers is superior to the abacus method of reckoning. This Arabian

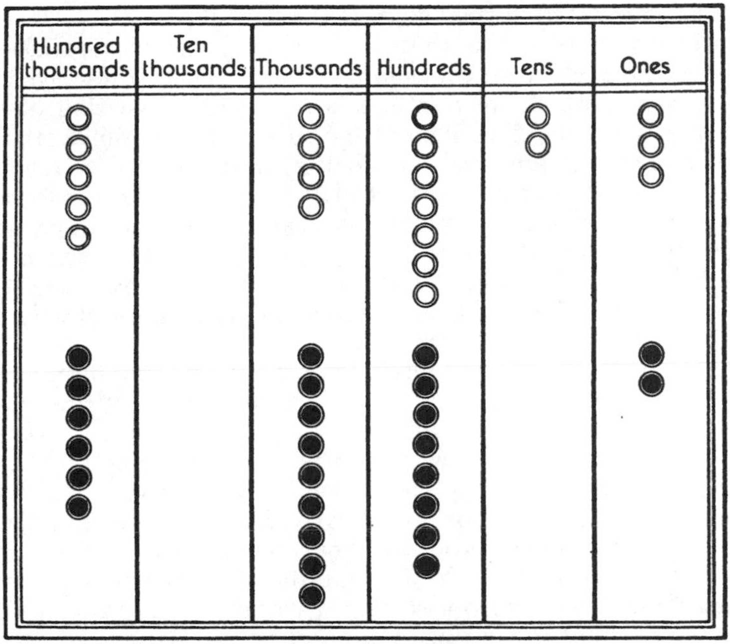

Hundred thousands	Ten thousands	Thousands	Hundreds	Tens	Ones

FIG. 1. The Abacus.

mathematician came from Khorassan and later lived in Baghdad. Between A.D. 800 and 825 he wrote among other things a work in which he describes the foundations of reckoning with the Hindu or so-called Arabian system of figures, using place values. He knew all about zero and wrote it as a small circle. By devious ways, from the crusaders and from the Moorish universities of Toledo, Seville and Granada, the Arabian works in Latin translation came to be known to scholars of Western Europe. Amongst these works was al-Khwarizmi's on Arabic numerals.

No longer were the clumsy abacus counting boards necessary for reckoning. Now an almost magical system of numbers

allowed long and complicated calculations to be carried out
with complete assurance. All that was needed to accomplish
this was the ten digits from 0 to 9, a pen, a piece of paper and
a knowledge of the multiplication table. It is difficult for us
to imagine the feelings of the people who first realised that
they could throw away their counting boards and in future
do a complicated operation on a scrap of paper.

But we must leave these somewhat romantic realms of
history and return to the commonplace. This Arabic system
of numbers which we now use is a system of writing down
certain methods of reckoning by means of certain symbols.
These methods are used within the limits of the system in
such a way that they do a part of the work of thinking for us.
Thus they enable our minds to soar in regions which our
powers of imagination could not reach, or at least, in which we
could very easily wander aimlessly. So we must examine this
Arabic system, the so-called system of tens, more closely to
understand wherein its strength lies.

Before we go on to the next chapter the budding mathematician should be given a few practical hints. Write your
symbols down as neatly and as carefully as possible. Don't
tinker with them, don't write things down in a muddle, don't
scribble workings in the margin or in odd spaces. Further,
never, as a beginner or as an experienced mathematician,
leave out steps in the process of your mathematical operations
because you have done them in your head. The whole process
must be written down on paper. If you like you can work the
whole thing out in your head in order to exercise your mental
powers, but you had better make notes on it before you go to
bed and then next day write it neatly step by step, checking it
according to the rules.

CHAPTER II

THE SYSTEM OF TENS

How extraordinarily simple our system of reckoning is. Really it consists of very little more than the arranging of ten symbols. If we add to these a few connecting signs, like the sign for " plus," " minus," " multiplied by " and " divided by " ($+ - \times \div$), and finally the " equals " sign ($=$), we have almost the whole system at our command. There is one more important thing to be mentioned, the system of place values. By this we mean that a symbol has a value which depends on the place in the number in which it finds itself. Its value increases the further to the left it is placed. Take the number 3333 ; the extreme right-hand symbol's value is simply 3, the next to the left 30, the next 300, and so on. The value of a number increases tenfold as its position is moved one place from right to left. That is why we speak of the system of tens ; 10 is the foundation or base of it.

The Romans used a system of numbers without place values. This made reckoning a difficult business. Suppose we had to add the numbers CCCXLIX and MMCXXIV. The position of the " I " in the numbers does not tell us that its value is in the range between ten and ninety. It tells us something about the value of the last sign. " IV " means " take 1 away from 5." This makes matters very complicated and it is not surprising that reckoning with Roman numerals entailed the use of a counting board. For us who use the Arabic numerals it is comparatively simple to write down 349 under 2124 and add the sum quickly in our heads.

If we are to examine more closely this system of tens which we use nowadays we had better be introduced to a new idea which will simplify the examination for us. This is the idea of a " power." When we speak of raising a number to a power we really only mean that we multiply it by itself as often as is indicated by the little number attached to the figure. For example, if we see 5^4 printed in a book, this merely means that we are expected to multiply $5 \times 5 \times 5 \times 5$; 5^6 means $5 \times 5 \times 5 \times 5 \times 5 \times 5$; 10^3 means $10 \times 10 \times 10$. We speak of " 5 to the power of 4," " 5 to the power of 6," " 10 to the power of 3." A number raised to the power of 2, *e.g.*, 10^2, is usually said to be " squared." We do not normally write 5^1, or " the power of 1." We leave a number unadorned in this case. But we can write a number with the power 0, like this : $10^0, 25^0, 3^0$. All these have one value, namely 1. Any number raised to the power of 0 equals 1. The reason for this will be explained in a later chapter, it cannot be given here. In the meantime the fact must be accepted.

Anyone looking at these numbers raised to powers will soon see that if 10^4 means multiply 10 by itself 4 times, the little number 4 (which is called the index number because it indicates the number of times to multiply), represents the number of noughts in the answer. This is only true when the index number is attached to the number 10, because 10 is the base of our number system.

Let us examine in greater detail any number in this system, say 3206. If we were putting this on to an abacus or counting board we should have to split it up into

$$6 \text{ in the column } 0\text{—}9$$
$$0 \quad ,, \quad ,, \quad ,, \quad \text{for 10s.}$$
$$2 \quad ,, \quad ,, \quad ,, \quad \text{for 100s.}$$
$$\text{and } 3 \quad ,, \quad ,, \quad ,, \quad \text{for 1000s.}$$

But there must be some method of writing this down on paper so that we can see from it how each number is constructed, without having recourse to a counting board. It is not so very difficult to discover such a method. We can write the number 3206 thus,

$$6 \times 1 + 0 \times (10) + 2 \times (10 \times 10) + 3 \times (10 \times 10 \times 10),$$

noting as we do so that any number multiplied by 0 equals 0.

We already know that, when a number is multiplied by itself, a short way of writing this is to put a small index number against it to show that it is being raised to a power. Also, a dot is an accepted alternative sign for $\times$, the multiplication sign. Therefore our example above could be written

$$6 . 10^0 + 0 . 10^1 + 2 . 10^2 + 3 . 10^3.$$

(Reminder : 10^0 always equals 1.)

Now we can see on paper the inner structure of a number in our system, that is, in the scale of ten. We can see the values of the places in which the separate figures are positioned in a number, that is to say their " *place value*." Our example

3206, when written $6 . 10^0 + 0 . 10^1 + 2 . 10^2 + 3 . 10^3$, is called a " *series* " of powers of 10. The 6, 0, 2, 3 are called the " *coefficients* " of the powers. Do not be put off by these technical terms ; we have used this example in which to introduce them because they will be useful in the next chapter.

Before we finish this one, however, and at the risk of boring the reader, we will have another look at our number system. We can see first of all that the system of place values is maintained in the series. The index numbers of 10 follow each other, 10^0, 10^1, 10^2, 10^3 (10^3 being a larger quantity than any of the others) ; the size of the coefficients makes no difference for $9 . 10^0$ is always smaller than $2 . 10^1$ or even than $0 . 10^1$; for $9 . 10^0$ represents 9 only, while $2 . 10^1$ equals 20. Suppose we take the number 109 and write it as a further example of a series :—

$$9 . 10^0 + 0 . 10^1 + 1 . 10^2$$

or if we like it in the reverse order, as it makes no difference to the sum of the series,

$$1 . 10^2 + 0 . 10^1 + 9 . 10^0.$$

We see that even the nought has a place in the series ; it shows that the next greatest place value is going to be filled. Theoretically this kind of series could be continued as long as we like. There is no number which is not capable of being expressed in rising powers of 10, each qualified by a coefficient. A perfect number system really requires that each step in the system of place values, each raising of 10 to a further power, should have a name of its own like ten, hundred, thousand. In our system there is not a complete set of names. There are special words to denote only 10^1, 10^2, 10^3 and 10^6, that is ten, hundred, thousand and a million. Ten thousand and a hundred thousand are simply multiplication expressions. This irregularity in naming the steps probably has its origin in some practical requirements of man in past ages. In very early times it is likely that money and armies only needed numbers up to thousands. It is said that the wealth of Marco Polo first made the concept of " million " necessary. Large numbers like a billion (10^{12} in England, 10^9 in America) are often called astronomical numbers, though they are rarely used in exact science.

Before we leave this chapter introducing our number system in the scale of ten, there are probably some questions which have occurred to the reader and which we must try to answer. Why do we use the words " eleven " and " twelve," only to follow them with " thirteen," " fourteen," " fifteen," and so on ? What does the peculiar French word for 80, " quatre-vingt," mean ? There is no doubt that these upset the regular picture we have drawn of our system of tens. " Quatrevingt " (4 × 20) is very similar in construction to " forty " (4 × 10). " Eleven " and " twelve " look suspiciously like a continuation of the numbers one to ten. They are not apparently composite numbers. Why do not we say " oneten," " twoten," " thirteen " (3-ten), " fourteen " (4-ten), instead of eleven, twelve, etc. ? Why indeed is ten the base of our system ? Is there anything about the number ten which makes it to be preferred before all others ? Is this system based on ten something God-given, descended from heaven ? Or is the reason for our preference merely the fact that we have ten fingers and that our ancestors used to count on theirs ?

There is absolutely no logical reason for preferring a system based on ten to one based on any other number. In the course of history there have been systems based on 60, 50, 20 and 12. In the year 1690 Leibniz, the great mathematician and philosopher, described the most remarkable of all systems, the binary, which uses nothing but the two numbers 0 and 1. In modern times this number system has been found to be convenient for use in electronic computing machines. Finally, the puzzling " quatrevingt " is nothing but a relic of a Celtic system based on 20 (fingers plus toes !) which has slipped into the French language.

CHAPTER III

Now that we understand the structure of our system of tens let us try to work out for ourselves another system based not on 10 but on a different number. We will choose one that is smaller than 10 for our first example. Let it be 6. This is called the *base*. We will be careful to construct our new system exactly on the lines of the system of tens and then we will see how far we get with it. In the first place we can say quite simply that, as the system based on 10 needed ten separate number symbols, 0, 1, 2, 3, 4, 5, 6, 7, 8 and 9, then a system based on 6 must need six symbols, 0, 1, 2, 3, 4 and 5. How are we to write the numbers six, seven, eight and nine in this new system ? We must refer back to our series of numbers raised to a power. The base raised to the first power was written 10^1, or just plainly 10. It was therefore the first number to be written with two number symbols, 1 and 0. In our new system which we are constructing on the base 6, let us write this base down as 10, remembering always that now it represents 6. This looks somewhat confusing at first sight because we are so accustomed to the sacred system of tens ! But as soon as we write down the first twenty numbers according to a system based on 6, we shall understand it better. For purposes of comparison we will write these twenty numbers first in the system based on ten and, immediately below, the same numbers in the system based on 6 :—

1 2 3 4 5 6 7 8 9 10 11 12 13 14 15 16 17 18 19 20
1 2 3 4 5 10 11 12 13 14 15 20 21 22 23 24 25 30 31 32

We will next look at our new system of numbers at its various levels. Just as we wrote 10^0 for 1, 10^1 for 10, 10^2 for 100, 10^3 for 1000, so we can write 6^0, 6^1, 6^2, 6^3 and so on, to represent the values 1, 6, 36, 216. . . . We must not forget that we are still using the scale (or system) of 10 here, and furthermore, we must never forget that the symbol 6 does not exist in our system based on the number six. To make matters clearer to ourselves we will in future write all examples in the scale of six in italics, all examples in the scale of ten in normal type.

So we can summarise the structure of our two systems in this way :—

Scale of 10
$10^0 = 1$
$10^1 = 10$
$10^2 = 100$
$10^3 = 1000$

Scale of 6
$10^0 = 1$
$10^1 = 10$
$10^2 = 100$
$10^3 = 1000$

Scale of 6 written in terms of scale of 10

$$10^0 = 1$$
$$10^1 = 6$$
$$10^2 = 36$$
$$10^3 = 216$$

If we returned to the use of a counting board or abacus and dealt in a system based on six instead of ten, the only difference would be that our columns would now be filled when they contained 5 counters or beads, the sixth counter having to go into and begin the next column.

Let us now take a number in the system of 6 and write it out as a series, still in terms of the system of 10 :—

$$2 \cdot 6^0 + 4 \cdot 6^1 + 0 \cdot 6^2 + 3 \cdot 6^3 + 5 \cdot 6^4$$

This amounts to

$$2 \times 1 + 4 \times 6 + 0 \times 36 + 3 \times 216 + 5 \times 1296$$

The answer is 7154. (It should be noted that even when writing in terms of the system of 10, the coefficient can never be greater than 5.)

Now let us take the same number and write it down in the scale of 6. All we have to do is to set the coefficients next to one another, beginning with the highest power and following on in descending order. Our number now looks like this :—

$$53042,$$

that is to say, it is merely the series

$$2 \cdot 6^0 + 4 \cdot 6^1 + 0 \cdot 6^2 + 3 \cdot 6^3 + 5 \cdot 6^4$$

We can check to see whether *53042* in the scale of six does really equal 7154 in the scale of ten, thus :—

$$7154 \text{ (scale 10)} = 4 \cdot 10^0 + 5 \cdot 10^1 + 1 \cdot 10^2 + 7 \cdot 10^3$$
$$53042 \text{ (scale 6)} = 2 \cdot 6^0 + 4 \cdot 6^1 + 0 \cdot 6^2 + 3 \cdot 6^3 + 5 \cdot 6^4$$

or $$4 \times 1 + 5 \times 10 + 1 \times 100 + 7 \times 1000$$

is the same as

$$2 \times 1 + 4 \times 6 + 0 \times 36 + 3 \times 216 + 5 \times 1296$$

Both series add up to the same number ; they both yield the result, written in the scale of 10, 7154.

If we leave the scale of 10 and write our number *53042* (scale 6) in a series in its own system we see that

$$53042 = 2 . 10^0 + 4 . 10^1 + 0 . 10^2 + 3 . 10^3 + 5 . 10^4$$

In this series the *10* is no longer the number 10 of the system of tens. It represents a 6 in the system of tens.

We can now attempt something further ; we can find out if it is possible to use this system of sixes in operations such as addition and multiplication. Before we can proceed however we need the help of just one thing, namely a multiplication table in the scale of six. At first sight this table looks rather crazy, but after a little pondering it soon appears quite sensible. We have to remember that there are only 6 number symbols to manipulate. Here is the crazy looking table in the scale of 6 :—

$$1 \times 1 = 1 \quad 2 \times 1 = 2 \quad 3 \times 1 = 3 \quad 4 \times 1 = 4 \quad 5 \times 1 = 5$$
$$1 \times 2 = 2 \quad 2 \times 2 = 4 \quad 3 \times 2 = 10 \quad 4 \times 2 = 12 \quad 5 \times 2 = 14$$
$$1 \times 3 = 3 \quad 2 \times 3 = 10 \quad 3 \times 3 = 13 \quad 4 \times 3 = 20 \quad 5 \times 3 = 23$$
$$1 \times 4 = 4 \quad 2 \times 4 = 12 \quad 3 \times 4 = 20 \quad 4 \times 4 = 24 \quad 5 \times 4 = 32$$
$$1 \times 5 = 5 \quad 2 \times 5 = 14 \quad 3 \times 5 = 23 \quad 4 \times 5 = 32 \quad 5 \times 5 = 41$$

We will now add, subtract, multiply and divide as if we had never heard of the system of tens.

First let us add *4325* and *5041*.

$$\begin{array}{r} 4325 \\ + \ 5041 \\ \hline 13410 \end{array}$$

Whenever two numbers added together come to six, we have to express this six as *10*. Thus *1* and *5* make *10* ; and so, in the example we write down *0* and carry *1* to the next column ; *4* plus this *1* are *5*, plus *2* makes *11*. We write down *1* and carry *1* ; *0* plus this *1* is *1*, plus *3* makes *4* ; *5* and *4* makes *13*. Of course we ought not to say " ten, eleven, thirteen," but rather " six, one six, three six." The main difficulty is a verbal one. Once we have words to describe the "step numbers" ten, twenty, thirty, hundred, thousand, etc., any system of counting is just as easy to manipulate as the system based on 10.

Now we will do a subtraction, still in the scale of 6 :—

$$5201$$
$$-\,3544$$
$$\overline{1213}$$

Let us see how this result is achieved. We will examine each stage of the subtraction in detail. First, *4* from *1* cannot be subtracted so we borrow from the next column ; we borrow *6* as we are working in the scale of six. Therefore we say " *4* from *6 + 1* is *3*." Secondly, in the next column, remembering that we have borrowed from it, we have to take *4* from *5*. This leaves *1*. Next we have to take *5* from *1* (*1* having been borrowed from the *2*). We cannot do this unaided ; we again borrow *6*. So *5* from *6 + 1* is *2*. Lastly we have to subtract the final column in which *3* is to be taken from *4* (because we borrowed *1* from the *5*) to leave *1*. That is how we arrive at our answer *1213*.

Now to multiplication. Let us multiply *3425* by *31* and make use of our multiplication table in doing so :—

$$3425$$
$$31$$
$$\overline{}$$
$$151230$$
$$3425$$
$$\overline{}$$
$$155055$$

It is just a matter of using the 3 times and 1 times table of the multiplication table in the scale of 6. We will check the answer by translating the sum into the scale of 10.

3425 (scale 6) $= 5 \cdot 6^0 + 2 \cdot 6^1 + 4 \cdot 6^2 + 3 \cdot 6^3$
  (scale 10) $=\quad 5 \;+\; 12 \;+\; 144 + 648 = 809$

31 (scale 6) $= 1 \cdot 6^0 + 3 \cdot 6^1$
  (scale 10) $=\quad 1 \;+\; 18 = 19.$

Thus the multiplication written in the scale of 10 is

$$809$$
$$19$$

$$8090$$
$$7281$$

$$15371$$

If we have calculated correctly, and if moreover our assertion is true that the laws which govern operations in the scale of 10 are equally applicable to the scale of 6, then 15371 (scale 10) must equal *155055* (scale 6). If we express these numbers as series they become

$$1 \cdot 10^0 + 7 \cdot 10^1 + 3 \cdot 10^2 + 5 \cdot 10^3 + 1 \cdot 10^4$$
and $$5 \cdot 6^0 + 5 \cdot 6^1 + 0 \cdot 6^2 + 5 \cdot 6^3 + 5 \cdot 6^4 + 1 \cdot 6^5.$$

If you care to work these sums out you will find that the two series are equal.

Now comes a division sum :—

$$425)2004013(2413$$
$$1254$$

$$3100$$
$$2552$$

$$1041$$
$$425$$

$$2123$$
$$2123$$

$$\cdots\cdot$$

The sum has " come out " exactly, as we say. Of course we have had to bear in mind all the time when doing the division that we are working in the scale of 6, from the very beginning when we made the first estimate which always has to be done when dividing. We always have to ask ourselves, roughly how many times the divisor will go into the dividend. How often does *425* go into the first group of figures in our dividend, *2004* ? If we were working in the scale of 10 we might have tried 4. But we must bear in mind that the *20* in the scale of 6 has the value 12 in the scale of 10, while 4 has the same value in both scales. As a *0* follows the *20* in the dividend and a *2* follows the *4* in the divisor, it is as though we were dividing 120 (scale 10) by 42. So we had better begin by trying *2*. To estimate the second place in the quotient (answer) we will divide *31* by *4*. But *31* is the same thing as 19 in the scale of 10. So we can write down *4* as an estimate. And so we proceed. Let it be noted that we can and must use an appropriate multiplication table when attempting these estimates, be it in the scale of 10 or any other number. If we use the multiplication table in the scale of 6 which we have already been given, it is unnecessary to work out each step in the estimation process in the scale of 10.

Now let us see if we can check the answer of this division sum. We could of course translate the whole into the system of tens in which we naturally feel more at home, as we did in the multiplication check. But we might try a different method this time. Whenever we are not sure whether the answer to a division sum is correct or not, we always multiply the divisor and the quotient together and see whether the result equals the dividend. If it does then the quotient is correct.

$$\text{Dividend} \div \text{Divisor} = \text{Quotient}$$
and
$$\text{Divisor} \times \text{Quotient} = \text{Dividend}.$$

As we do not want to refer to the scale of 10 any more than necessary when working in the scale of 6, let us multiply the numbers *2413* and *425*.

$$
\begin{array}{r}
2413 \\
425 \\
\hline
1450000 \\
52300 \\
21313 \\
\hline
2004013 \\
\end{array}
$$

We should be satisfied. The check has proved our answer to be correct ; the dividend has emerged accurately from the multiplication check.

Someone may at this point observe that we did not in actual fact multiply the divisor by the quotient to obtain the dividend. What we did was to multiply the quotient by the divisor. That is a correct observation to make, but it does not really make any difference whether you multiply 5 by 4 or 4 by 5. The answer is the same. However, it does give us an opportunity to digress for a moment.

Addition and multiplication are operations which do the work of building up. Their work is to unite or to increase ; they synthesise. Subtraction and division on the other hand are operations which do the work of splitting up and reducing. They analyse into parts. It is clear, or perhaps we should be more cautious and say it is probable, that the group which synthesises and that which analyses each has common characteristics. We will not go into the matter more deeply at this point but we will content ourselves with the observation that addition and multiplication, as opposed to subtraction and division, possess a very important common characteristic which everybody knows. The terms of addition and multiplication sums are interchangeable. It makes no difference to the result whether we add $5 + 4 + 7$ or $4 + 5 + 7$ or $7 + 4 + 5$ or any other combination of these three numbers. The same is true of $4 \times 5 \times 7$. The rule therefore is that operations involving synthesis (multiplication and addition) obey the " *Commutative Law* " ; that is, the various parts are interchangeable and the additions and multiplications can be made in any order. On no account can this law be applied to analytical operations, which by the way at this stage deal only with two terms. To take 4 from 5 is an entirely different matter from taking 5 from 4. Similarly, 12 divided by 3 is different from 3 divided by 12. It may look as though this digression were an unnecessary statement of the obvious. There are however some more advanced synthetic and analytical operations in which all is not quite so obvious. When we come to these it will be seen that the obvious remarks have been worth making.

We must get back to our number systems. After our experience with the scale of 6 we can assume that we understand the workings of number systems with a base less than 10, but we have not proved that a system with a base greater than 10 is suitable for place values. We had better not choose a base as large as 50. It would be possible to do so, but the size of the multiplication table and the effort required to memorise it would soon make us giddy and we should lose any general idea of the system. What is more, we know that we need as many separate number symbols as the base has

units—in this case 50. Where are we to get these symbols
from ? It would take days to invent them and then to learn
them.

Let us be satisfied with a modest number greater than 10,
namely 13. We can demonstrate with this number that a
" *prime number*," that is one not divisible by any other, is
also suitable to be chosen as a base. It has been quite
seriously suggested more than once in the past that the
system of tens should be abandoned in favour of twelve
because our base, 10, is only divisible by 2 and 5, whereas 12
is divisible by 2, 3 and 4. There is no doubt that the system
based on 12 would have untold advantages for some monetary
systems and for weights and measures, apart from the fact
that the hour, minute and second divisions of the day as well
as the angle divisions of the circle would fit into the system
of 12 very easily. Arguments against this system are mostly
founded upon observations of nature, namely that the number
of our fingers and other parts of the body is, generally speaking,
based on 5 or 2 (fingers, toes, ears, eyes, arms, legs). What is
more, the whole metric system (system based on 10) with all
its extensions into the realm of decimals is closely connected
with the earth, as the metre has been defined ever since the
French Revolution as the 10-millionth part of the distance
between the North Pole and the Equator. Other quantities
of measurement, such as the litre and kilogram, are linked
with the metre. And finally, by a curious coincidence, the
most important measurement of the universe, the so-called
speed of light, is almost exactly 300 million metres per
second.

So there is not much prospect of having to learn a new
system of numbers in the foreseeable future ! We will do some
work in the scale of 13 not so much for practical reasons as
for the demonstration of an underlying principle. Let us
write down the first 30 numbers of the system of 10 and below
write them in the system of 13 :—

Scale 10	1	2	3	4	5	6	7	8	9	10	11	12	13	14	15
„ 13	1	2	3	4	5	6	7	8	9	A	B	C	10	11	12

Scale 10	16	17	18	19	20	21	22	23	24	25	26	27	28	29	30
„ 13	13	14	15	16	17	18	19	1A	1B	1C	20	21	22	23	24

As can be seen, we have brought into use the three capital

letters A, B and C, because in a system based on 13 we need
twelve single-number symbols as well as the 0.   When we
constructed the system of 6, four symbols of the scale of 10
(6, 7, 8 and 9) did not occur.   We simply left them out.   Now
it is just the other way round.   The system of 10 excludes the
three numbers A, B and C.

We could at this stage write out a multiplication table in
the scale of 13 in which, for example, $5 \times 8 = 31$ and
$7 \times 7 = 3A$, but we will leave this task to the reader who
wants to penetrate deeper into the construction of number
systems.   Suffice it to say that $A \times B = 86$.

For all that, we must show in some way or other that our
system of 13 works.   Let us multiply 92B by A7 which, if we
had to say it aloud, might sound like " ninehundredand-
twentyBe multiplied by Atyseven " !

$$92B$$
$$A7$$
$$\overline{}$$
$$71260$$
$$4C6C$$
$$\overline{}$$
$$761CC$$

Putting this into words, we have multiplied thus : A times B
is 86 ; put down 6 and carry 8 ; A times 2 is 17, plus 8 is 22 ;
put down 2 and carry 2 ; A times 9 is 6C, plus 2 is 71.   Next,
7 times B is 5C ; put down C and carry 5 ; 7 times 2 is 11,
plus 5 is 16 ; put down 6 and carry 1 ; 7 times 9 is 4B, plus 1
is 4C.   Now adding the two lines together, in the first place
$C + 0 = C$;   in   the   second $6 + 6 = C$;   in   the   third
$C + 2 = 11$, write   down   1   and   carry   1 ;   in   the   fourth
$4 + 1 = 5$ plus 1 is 6 ; in the fifth place 7.   Therefore the
answer, in the system of 13, is 761CC.

In order to save time we will commit something of a crime
and do the check in the scale of 10 by expressing the sum as a
series.

92B (scale 13) $= B \cdot 13^0 + 2 \cdot 13^1 + 9 \cdot 13^2$
      (scale 10) $= 11 \times 1 + 2 \times 13 + 9 \times 169 = 1558$

A7 (scale 13) $= 7 \cdot 13^0 + A \cdot 13^1$
      (scale 10) $= 7 \times 1 + 10 \times 13 \qquad\qquad = 137$

Now we will multiply in the system of ten :—

$$
\begin{array}{r}
1558 \\
137 \\
\hline
155800 \\
46740 \\
10906 \\
\hline
213446
\end{array}
$$

The result of this multiplication in the system of 13 was 761CC. This number should be equal to 213446 in the system of 10.

Therefore the series $C . 13^0 + C . 13^1 + 1 . 13^2 + 6 . 13^3 + 7 . 13^4$ should add up to 213446 (scale 10). Let us see if it does :—

$$
\begin{aligned}
&12 \times 1 + 12 \times 13 + 1 \times 169 + 6 \times 2197 + 7 \times 28561 \\
&= 12 \quad + 156 \quad\;\; + 169 \quad\;\;\; + 13182 \quad\;\; + 199927 \\
&= 213446.
\end{aligned}
$$

So we get the result we have been expecting and have shown that a system with a base greater than 10 obeys the rules of any system of place values. It should be noted that a mathematician does not regard this as a " proof." He would at most call it a verification. At this stage however let it be proof enough for us.

What have we learnt about all these number systems of the place value type ? Each is an infallible automatic system of thinking and reckoning. The construction of each is the same :—A base—as many figures (including 0) as the base has units—place value, each coefficient being thought of as multiplied by whatever power of the base is allotted to it. The first place has the power 0, each following place has a power increased by 1. When the powers are worked out we call the results step numbers. To constitute a practical system of reckoning, each step number should have a name of its own ; especially is this true of the lower ranges. Every system has one, two, three and more figure numbers. A number always has one more figure in it than the value of its power indicates, that is to say $10^3 = 10 \times 10 \times 10 = 1000$. Further, the same rules for adding, subtracting, multiplying and dividing apply to any number system with place values.

Before we pass to the last result of our examination of number systems, let it be noted that a simple method of calculating is not the only pre-requisite of a trouble-free " *written* " system of reckoning. The Arabic system of place values is actually needed before it is possible to invent such well-known calculating machines as the cash-register in shops and the meter in taxis. Calculating machines proper, such as are used in banks or accounting and technical offices, are based on the theory of place values too, as we mentioned earlier.

There are one or two other fundamental ideas of mathematics which emerge as the result of our labours, as well as the idea of an automatic purposeful system of reckoning. These are the ideas of generality, similarity and invariance. As we do not want to become involved in mathematical philosophy we will illustrate these highly abstract ideas by examples taken from our previous work.

We began with our system of ten which at first we imagined to be of divine origin. Then we saw that it was just one of innumerable possible systems. This brought us to the general idea of a number system with a means of writing place values. For this system, which can have any number we care to choose for its base, we formulated some general rules which are not only valid for one case but are applicable to all such systems. Therefore these rules are universal. But the systems have to be alike, similar, in form too. And constancy of form, or invariance, means that a series of rules applied to one system is applicable to all other similar systems. The systems of 10, of 6, of 13, and all the countless other systems of place value are alike in form. That is why, for example, the rules for multiplying apply to them all. Systems of place value are similar so far as multiplication is concerned ; they are " *invariant* " as regards multiplication. We always carry out the process in the same way and obtain the same result whatever the base of the system may be. So we could immediately change any calculating machine from a system of 6 to one of 13 without altering the principle on which it was constructed, simply by exchanging a few parts. It would then obediently deliver the result, expressed in the other system.

Some of us may however be assailed by doubts. We mentioned the existence of a system based on 2 and discovered that the whole multiplication table of this system which contains only two number symbols 1 and 0, consists of one line, $1 \times 1 = 1$. This may be very attractive to any learner, but it is rather confusing for we stated that we can work any system by applying the rules. Yet how are we going to multiply large numbers if all we know is that $1 \times 1 = 1$? The answer is that there is hardly any multiplication to do at all— it is almost all addition. The reader may like to verify that the multiplication sum

$$1101 \times 101 \text{ (scale of 2)}$$

given below is merely a rather long-winded way of multiplying 13 by 5 (scale of 10).

$$
\begin{array}{r}
1101 \\
101 \\
\hline
110100 \\
1101 \\
\hline
1000001
\end{array}
$$

Then there is a second problem. As soon as we try to work out according to our present knowledge how many 2-figure, 3-, 4- or 10-figure numbers there are in any chosen system, we run up against all kinds of difficulties.

It seems then that we shall be obliged to continue still further our researches into numbers before we can finally turn aside from the " theory of numbers " and interest ourselves in algebra, that is, calculation with generalised numbers.

CHAPTER IV

SYMBOLS AND COMMANDS

ONE of the many questions which may have occurred to us when we were examining the structure of number systems may well have been this—how is it that, starting with a limited number of symbols, we are able to build up any numeral system in which numbers can be written without limit ? Can this be done even when using only 0 and 1 as in the scale of 2 ? We have known the Arabic system of 10 since our childhood ; it is the system in which we shall in future be doing virtually all our work. For that reason we dare not assume that we know all about it, although we have been familiar with it for so long. Take for instance a number written like this—3 !. What does it mean ? When we can answer that we shall have gained further knowledge of our number system and shall be able to answer the first questions we posed in this chapter.

What is the meaning of the exclamation mark—3 ! ? It is an order, a command to us to undertake a certain action. What kind of an action ? We are commanded to multiply it in a certain way. Of course we are at liberty to put this command sign, the exclamation mark, after any number we like, not only 3.

Before we examine this special command sign in greater detail let us take a look at the different kinds of mathematical commands and their uses. We have been obeying quite a number of such commands already without noticing them because we have been used to do so from our very first lessons. We have shown that single figures and numbers made up from these figures are signs or symbols standing for certain quantities. And we have spoken of a system, an ingenious method of calculating. There is something more besides all this, the commands. It is only when we understand these commands that we can develop our system of numbers and our methods of calculating from the single figures we started with. If we were to call these methods of calculating " operations " we could go on to speak of " operational commands " which might be called " operational signs " when written. But we shall continue to speak of " commands " whereby is meant, of course, a mathematical command to undertake some piece of calculation.

It is always difficult to carry out commands correctly and precisely in any new field of activity ; so in mathematics too the greatest difficulty for the beginner lies in understanding and accurately carrying out the commands. Nine-tenths of mathematical skill however consists in the understanding of this " mathematical discipline."

Let us, as ever, begin with the simplest case. We have already actually used the command and carried out the operation of addition. $5 + 4 = 9$. What does this statement in figures mean in words ? It means that we are asked to take five units and count four more units on. After a short pause in which to carry out this operation, a sign " $=$," the equals sign, is written down. This indicates that the command has been faithfully obeyed. Finally, when this has been done a new symbol appears to the right of the equals sign, namely the 9. The job is done, the operation completed.

Subtraction too is a command, and multiplication and division also. Everyone knows how complicated a mathematical command can be to carry out—for example, the division of a number running to several places, in the scale of 13. The fact that we are asked to use a certain system of numbers is also a mathematical command ; so is the use of indices.

There are therefore many examples of this idea of a mathematical command for us to note. Let us go back now to our example at the beginning of the chapter in which the exclamation mark was used as a command sign. What does 3 ! mean ? What are we told by the sign to do ? It means that we have to calculate the " factorial " of 3. We are told to take the figure 1, multiply it by 2, multiply the answer by 3. That makes 6. So 3 ! = 6. Let us take 5 !. Begin with 1, multiply by 2, multiply the answer by 3 and continue thus until you have multiplied by the number in front of the exclamation mark. If we read the command 1 ! there is nothing we need do ; it is like raising a number to the power of 1. Here is a table setting out the factorial numbers from 1 ! to 16 !.

$1! = 1$	1
$2! = 1.2$	2
$3! = 1.2.3$	6
$4! = 1.2.3.4$	24
$5! = 1.2.3.4.5$	120
$6! = 1.2.3.4.5.6$	720
$7! = 1.2.3.4.5.6.7$	5,040
$8! = 1.2.3.4.5.6.7.8$	40,320
$9! = 1.2.3.4.5.6.7.8.9$	362,880
$10! = 1.2.3.4.5.6.7.8.9.10$	3,628,800
$11! = 1.2.3.4.5.6.7.8.9.10.11$	39,916,800
$12! = 1.2.3.4.5.6.7.8.9.10.11.12$	479,001,600
$13! = 1.2.3.4.5.6.7.8.9.10.11.12.13$	6,227,020,800
$14! = 1.2.3.4.5.6.7.8.9.10.11.12.13.14$	87,178,291,200
$15! = 1.2.3.4.5.6.7.8.9.10.11.12.13.14.15$	1,307,674,368,000
$16! = 1.2.3.4.5.6.7.8.9.10.11.12.13.14.15.16$	20,922,789,888,000

It can be seen that the command sign ! leads very quickly to extraordinary consequences. The beginning is harmless enough but the results of the command rise suddenly and keep on increasing until they soon reach fantastic numbers. Factorial 100 is an enormous number of 158 figures.

Without going into the matter in detail, it can be seen at a glance that the series of " factors " bears a certain resemblance to powers. The difference is that when raising a number to a power we always use the same number to multiply by. But in factorials the multiplier increases step by step. For purposes of comparison let us take a low number and raise it to a power, remembering the story of the Persian mathematician Sessa Ebn Daher, the inventor of chess. The Indian king Shehram told him to choose anything he wanted as a gift. The mathematician, with an innocent look, said : " Mighty King, my wish is extremely modest. I should like to be rewarded in grains of corn. May I receive as much corn as would accumulate on the last square of the chess-board, if one grain is on the first square and twice the number of grains as are on the preceding square on all the following squares ? " The king laughed loudly and granted the mathematician his wish. He was quite sure that the man would not receive sufficient corn to make a loaf of bread. But he was quickly disillusioned for the number of grains of corn worked out to be $1 \times 2 = 2$, $2 \times 2 = 4$, $4 \times 2 = 8$, $8 \times 2 = 16$, $16 \times 2 = 32$, $32 \times 2 = 64$, $64 \times 2 = 128$, $128 \times 2 = 256$ or $(1 \times 2 \times 2 \times 2 \times 2 \ldots \ldots)$ and so on until the last square of the chessboard was reached. As a chessboard has 64 squares the total of grains of corn was 2^{63} because there was only one grain on the first square. But 2^{63} equals

$$9,223,372,036,854,775,808.$$

To realise what this number means it should be noted that the corn on the last square of the chessboard would need for storage space a cube whose sides were almost 5 miles long, taking the average volume of one grain of corn as 0·0028 cubic inches. (This is a figure derived from an experiment in which it was found that the number of grains of corn in a litre measure was 22,000.) It would weigh 16,000,000,000 tons.

It should also be noted that the world's wheat harvest for the years 1927 to 1931 was an average per year of about 6,000,000 tons. So the king in the story would have had to give the mathematician about 2,600 world wheat harvests of the twentieth century, produced as they are with the aid of tractors and artificial manures, if he wanted to keep his promise.

The number of grains required by the inventor of chess needed a 19-figure number for its expression, but the reader will realise how small this is compared with the 158-figure number of 100 !.

CHAPTER V

ARRANGEMENTS

HAVING digressed in order to show how an apparently simple mathematical command can lead to an enormous growth in number size, let us see how factorial numbers are used ; that is to say, in what mathematical field we can use the idea $1 \times 2 \times 3 \times 4$, etc., or the 3 ! or 4 ! etc. We shall use factorials in calculating the number of different ways in which things can be arranged.

It is not our intention to confuse the reader with a series of definitions. This would be the usual text-book approach. We prefer to proceed by explaining the subject in terms of what interests the reader and what he is already familiar with. But mathematics is not in itself an experimental subject like science, which is based on experience. It is purely intellectual, spun out of the brain without any need for experiment. Its results can neither be proved or disproved by experience ; proof can only be obtained by ensuring the accuracy of its logical operations. It is what philosophers call an *a priori* subject.

Let us therefore work with mathematical material with which we are already familiar. Let us ask ourselves once again how it is possible to make up all the numbers in the world out of the ten number symbols of our system of tens. Anybody looking at the numbers 123, 132, 213, 231, 312, 321, who has any idea of the meaning of the word " arrange " would say that these various numbers are arrangements of the single-number symbols 1, 2 and 3. That is quite correct. All numbers in the systems we have discussed are created by arrangement. We change the figures round, construct the numbers by altering the order of the symbols and thus give every number a different look, as it were. What interests us now is simply the possibility of distinguishing between numbers by comparing their external " appearance." After considering the matter more closely however, we discover a great difference in the way in which they can be arranged. It is not only a matter of changing the figures round ; the same figure can appear more than once in a number. In fact the same figure can appear as often as there are places in the number to be filled. Indeed, as soon as a number contains more than ten places figures are bound to appear more than once in a number, as we only have 10 number symbols to deal with.

Our next problem will be to consider arrangements of numbers (and other objects) to discover, if we can, how such arrangements behave. What rules do they obey? Can we count the number of possible arrangements? The answers to these questions deserve a chapter on their own.

CHAPTER VI

ONCE upon a time there was a good honest family consisting of a father, mother and twelve fine healthy children. The family was quietly sitting around the dinner table one day. Suddenly the youngest boy exclaimed that he never got his fair share of soup but only the remains because his place at table was unfavourably situated for the dishing out of soup. The members of his family got on well together and were in the habit of settling their differences by compromise. ' In short, they decided to change their places at table every day since they could not persuade the maid-servant to alter her practice of always serving the places in the same order. This decision gave rise to general discussion and they tried to estimate the amount of time it would take to exhaust all the possible ways of arranging their places at table. One boy thought it would take a few days. One of the girls, on the other hand, thought it would be safer to say some weeks. At last they agreed upon one year. The eldest boy was then heard to say that there was a formula for working it out. " What do you consider our case is, mathematically speaking ? " asked his father in order to test him. The eldest boy thought it over and said, " As we are concerned in this instance with the order of sitting at table, it makes a difference whether Jane sits next to Henry or Henry next to Jane. They are two quite distinct cases. What is more, no groups of people must be allowed always to sit together. Each one of us 14 people is put each time in a different order. It is exactly the same as if I were to put 14 things in all possible arrangements. This way of changing things about is called ' permutation,' and the formula for it runs as follows : if we are changing 14 things about, the number of changes is factorial 14 (14!), and similarly for any other number of things." His father was satisfied with this answer. The children fetched pencils and paper in the interval between courses and the older children began to calculate. How big is factorial 14 ? This frightful answer was obtained :—

87,178,291,200.

How long would it take to exhaust this number of possible orders at table ? The year has 365 days, so the children divided by 365 and they discovered that if they only changed places once a day it would take them almost 239 million years to exhaust the possibilities ! If they changed places twice a day it would take them more than 119 million years, and even by changing at all four meals every day they would still need almost 60 million years.

This example demonstrates both the really devilish multiplicity of possible interchanges and the first type of arrangement. Now we have acquired some fresh material which we will examine systematically.

First let us allay the fears of some of our readers. The enthusiasm of our intelligent family having been damped by the results of permutation, they hit upon a simpler way of dealing with the justifiable complaint of the youngest boy. The maid-servant was asked always to begin serving at the same place at the table, regardless of its occupant. The whole family however moved on one place each day, in a clockwise direction around the table. As far as their relationship to each other was concerned the order of sitting at table was undisturbed. Each member of the family kept his usual neighbours. If the table is considered as a source of food, however, then the arrangement by which food is obtained at it changes every day. This solution to the problem provided that each member of the family received his food first once every 14 days.

Looked at from a mathematical point of view, this arrangement results in 14 separate permutations, some of which we will now give :—

1,   2,   3,   4,   5,   6,   7,   8,   9,   10,   11,   12,   13,   14   (1st day)
2,   3;   4,   5,   6,   7,   8,   9,   10,   11,   12,   13,   14,   1   (2nd day)
3,   4,   5,   6,   7,   8,   9,   10,   11,   12,   13,   14,   1,   2   (3rd day)

and so on

14,   1,   2,   3,   4,   5,   6,   7,   8,   9,   10,   11,   12,   13   (14th day)
1,   2,   3,   4,   5,   6,   7,   8,   9,   10,   11,   12,   13,   14   (15th day)

Be it noted however that these 14 permutations (the first and the fifteenth are the same) have been picked out artificially from the total of 87,178,291,200 possibilities, according to another principle, namely the principle of cyclical order ; for an additional factor has to be borne in mind, that the relative order of seating at table remains unchanged.

In this example we used numbers to represent the changing places at table. The numbers were in the nature of indicators or labels. They were used just as they are for seats in a theatre. They can be used like any other distinguishing mark to show in what order a series of objects is arranged. The numbers tell us nothing about the respective sizes of the objects ; they are not meant to. But they do tell us what the relative position of the objects is when we examine a row of them for this purpose. We can say that object 2 is " higher " than object 1, or that it is the " higher element." We mean that object 2 comes after object 1 and we are not making any statement about the height or size of the objects. It would probably be less confusing if we used letters of the alphabet instead of numbers as indicators, for numbers do carry with them a suggestion of size which is not appropriate here. We can equally well say that letter " d " is " higher " than letter " a," meaning thereby that in an arrangement object d comes higher in or further along the series than object a.

This brings us to an idea which is common to all types of arrangements. We will call it the idea of " good order." A " good order " occurs if we permute in the following manner :—

abc, acb, bac, bca, cab, cba

or in numbers :—

123, 132, 213, 231, 312, 321.

By progressing in this manner so that the " lowest " element keeps its place as long as it can, no possible permutation can escape us ; we rise from the " lowest " to the " highest " permutation in good order. When the original permutation turns up back to front it is an indication that all possible permutations have been completed. In the lowest permutation, if we are dealing with numbers, the numbers run from the lowest to the highest, reading from left to right : a lower element always comes before a higher one. In the highest permutation the position is completely reversed and the higher element always stands in front of the lower. We can note in passing that if we are permuting numbers the lowest permutation will actually give the lowest number and the highest permutation the highest number, e.g., 123 and 321. Between these lie all the other numbers it is possible to make up with 1, 2 and 3 in increasing size if permuted in good order.

Let us return to our idea of indicators or labels and look

upon 123 and 321 as merely permutations of the indicators
1, 2 and 3. We want to see now how it is that the factorial
number, for instance 3!, gives us the number of permutations
which can be made with the given number of indicators. It
would be logical to begin with the simplest case. If we begin
with a single indicator or element the problem is to apply to
it our rule of good order. We have to write down the number 1
and arrange it in good order. This cannot take us long as it
is obvious that the number 1 can only be arranged in one way.
In other words, the number of permutations possible with
one element is 1 or, if we prefer it, factorial 1 which equals 1.

The next logical step is to deal with two elements, which
we can indicate by letters a and b. The only two possible
permutations are   ab   ba

If we refer back to our rule of good order we see that in the
second permutation the order of the first one has been completely reversed. Therefore all possible permutations have
been completed. We must try to link this now with the result
for one element. Notice that we keep a in the first position
and give to b all its possible permutations, namely one. Then
we keep b in the first position and give a all its possible permutations, again one. The total number of permutations is
therefore 2 ; we have 2 elements ; each element is capable
of only 1 permutation ; 2 elements are capable of 2 × 1
permutations or of factorial 2 permutations.

The permutations of 3 elements a, b and c can be written :—

$$abc \quad bac \quad cab$$
$$acb \quad bca \quad cba.$$

When we look at these arrangements we see that each element
in turn occupies the first position while the other two are
permuted ; the 3 elements in turn remain fixed while 2 are
permuted. Therefore the number of permutations of the
3 elements is $3 \times 2! = 3 \times 2 \times 1 = 3! = 6$.

With 4 elements we can write the following permutations :—

abc 1	bacd	cabd	dabc
abdc	badc	cadb	dacb
acbd	bcad	cbad	dbac
acdb	bcda	cbda	dbca
adbc	bdac	cdab	dcab
adcb	bdca	cdba	dcba.

All that need be said about these arrangements is that each of the 4 elements has been kept in the first position while the other 3 have been permuted in every possible way. The total number of permutations is therefore $4 \times 3! = 4 \times 3 \times 2 \times 1 = 24 = 4!$.

Provided that we use all the elements every time we permute, we can see from the examples in the previous paragraphs that the number of permutations will equal the " factorial " of the number of elements, *e.g.*, with 10 elements the permutations using all the elements every time will be 10! ; with 52 elements they will be 52! and so on.

It might so happen that in some cases we might be dealing with some elements which could be treated as though they were the same. Suppose we had to arrange in order 2 apples, 3 bananas and 1 cherry. Usually there is no need to distinguish between the first and the second apple or between the three bananas. In other words, the arrangement

Banana 1, banana 2, banana 3, apple 1, cherry, apple 2 is the same as

Banana 3, banana 1, banana 2, apple 2, cherry, apple 1, and so on.

For the sake of simplicity we will call apples *a*, bananas *b* and cherries *c*. Then the first permutation is

aabbbc

and the last   cbbbaa.

We will not dally at this point but will present the necessary formula without further ado. This is the case in which we are said to permute elements or things, some of which are alike. The rule is : first find the total number of permutations as though the things were all unlike ; secondly find the number of permutations for each group as though its elements were unlike ; then divide the total number of permutations by each of the group permutations in turn. For our example with apples, bananas and a cherry the total number of things was 6 ; if we pretend that these things are all different the total number of permutations is 6!. As they are actually *not* different we divide this total by 2! for the apples, 3! for the bananas and 1! for the cherry. The answer is therefore

$$\frac{6!}{2! \times 3! \times 1!} = \frac{6 \times 5 \times 4 \times 3 \times 2 \times 1}{2 \times 1 \times 3 \times 2 \times 1 \times 1} = 60.$$

This rule which applies when some elements are alike is really a more general rule for permutations than the simple factorial rule we first arrived at. Let us apply the more general rule to the case in which we permute five things. The total number of permutations we obtain by this rule is

$$\frac{5!}{1! \times 1! \times 1! \times 1! \times 1!}$$

as we have 5 groups, each containing 1. It is clear that, since $1! = 1$, the result we have just obtained is the same as that which our simple rule would have produced, namely $5!$. As the second rule gives all the results obtained by the first rule and more besides, it is said to be " more general."

It is worth while noting when applying this second rule that the number used with a factorial above the line equals the sum of the numbers used with factorials below the line ; for the total number of elements is the same whether they are counted in groups or not.

There are further problems of permutation which could be considered, but we have done enough to help us on our way with arrangements. Let us summarise what we have done in this chapter. We have stated that if we take a set of elements, then the number of changes we can make in their order is the number of permutations we can make of those elements. In other words, permutations are concerned with the mixing up of any given set of elements.

CHAPTER VII

COMBINATION

WE shall find the results of the last chapter very useful in considering a further kind of arrangement. At this point the unusually large family of 14 members puts in an appearance again. It is after dinner. They are considering some mild form of entertainment and a game of cards is suggested. Someone remembers the argument about the total number of possible seating arrangements at the dinner table and the question is asked—how long would it take for all the possibilities of a daily game of bridge to be exhausted ? Interest is centred on the number of ways in which they could make up a 4 rather than on the details of the game itself. Each game requires 4 players and every day the players are to be different. All 14 members of the family are to take part as required.

They are all shy of offering an opinion as to the length of time needed. It might take a million years again before the whole family had had its turn. They decide that it would be more sensible not to guess but to follow mathematical rules of arrangement. The mathematically-minded son declares at once that this time they are concerned with the particular form of arrangement known as " *combination*." Fourteen people, mathematically speaking, are the same as fourteen things or elements. The family is only concerned with the players making up different games of bridge, not with their order within the group or with their re-arrangement within it. A set of players made up of Mother, Father, Henry and Jane is the same as one made up of Father, Henry, Mother and Jane.

We can arrive at an answer to the problem of how many sets of 4 players there are in a family of 14 members in a few moments because it is easily obtained by using a so-called " *binomial coefficient*." This name has nothing to do with our example : it is taken from the binomial theorem from which it is derived and we do not propose to investigate this now but to accept for the time being the method of calculating combinations from it.

33

The number of elements is written down first, then a capital C (to stand for Combination) and finally the size of the group :

$$_{14}C_4$$

To work this out we must write

$$_{14}C_4 = \frac{14 \times 13 \times 12 \times 11}{1 \times 2 \times 3 \times 4} = 1{,}001.$$

So 1,001 games of bridge would have to be played before all 14 members of the family had combined in all possible ways to make up a game. Playing one game a day it would take them more than 3 years.

The members of the family are heartened by this and one of the girls puts a second question. " It so happens," she says, " that in our family there are 6 boys and 6 girls. I should like to know how many different pairs of dance-partners we could form. That must be a combination too, for there would be 12 things or elements to be arranged in 2 kinds of groups and it doesn't really matter if Jane dances with Henry or Henry with Jane." " You are quite right, Maggie," says the mathematical brother. " It is a combination of pairs. I might add that your little problem can be quite simply solved without using the combination formula. Each of the 6 girls dances with each of the 6 boys, that is to say 6 times. So there are 36 different dancing couples." " What is the use of your formula then ? " asks Jane. " This was a case which was easy to work out. I will show you how the formula you despise makes the calculation understandable and confirms the result. Well, all pairs made up from 12 elements would be written

$$_{12}C_2$$

But amongst these pairs would be some consisting of 2 brothers and some of 2 sisters, which are not the kind of pairs we want. We want proper dancing partners of opposite sexes in our pairs. So we must subtract the unwanted pairs. But these are also combinations made up of 6 elements when brothers are paired and the same number when sisters are ; that is to say, the number of paired brothers can be represented by

$$_6C_2$$

and the number of paired sisters is exactly the same. The calculation is really complete already.

$$_{12}C_2 - {_6}C_2 - {_6}C_2 = {_{12}}C_2 - 2 \times ({_6}C_2)$$

$$= \frac{12 \times 11}{1 \times 2} - 2 \times \frac{6 \times 5}{1 \times 2} = 66 - 30 = 36$$

That is the result we anticipated."

In our chapter on permutation we introduced the rule of " good order." Let us see how this works with combinations. Let us make up groups of 3 from 6 elements which we will indicate or label by the numbers 1, 2, 3, 4, 5 and 6. Here they are :—

123	135	234	256
124	136	235	345
125	145	236	346
126	146	245	356
134	156	246	456

We know that the last group of 3 has been achieved when we find one made up of the last 3 elements in good order, that is 4, 5, 6.

Suppose we want to know the number of combinations possible with 9 elements taken in pairs. Using the indicators 1, 2, 3 . . . 8, 9, the first pair would be 12 and the last 89. There is one more important rule to remember when dealing with indicators, namely that a " higher " number may never come before a " lower " one. By this rule we must not write 42 for the combination of 4 and 2 ; it should be 24.

For the sake of practice we will now combine in groups of 4 the elements a, b, c, d, e, f and g.

abcd	acde	adef	aefg	bcde	bdef	befg	cdef	cefg	defg
abce	acdf	adeg		bcdf	bdeg		cdeg		
abcf	acdg	adfg		bcdg	bdfg		cdfg		
abcg	acef			bcef					
abde	aceg			bceg					
abdf	acfg			bcfg					
abdg									
abef									
abeg									
abfg									

The method is obvious.

How can we determine the number of possible combinations of any particular number of elements in groups of a stated size ? Let us begin by finding the number of pairs we can make from 6 elements. To do this we must combine each element with each of the remaining 5. This gives 6 × 5 or 30 pairs, but in these pairs we are counting not only pairs like ab but also those like ba. So we have counted all pairs twice and must therefore divide 30 by 2. The value of $_6C_2$ is therefore $\dfrac{30}{2}$, which we will write as

$$\frac{6 \cdot 5}{2} \text{ or } \frac{6(6-1)}{1 \cdot 2}.$$

In the same way, if we want to find the number of combinations of 3 elements which we can make out of 6 elements, we have to be careful not to count each permutation as though it were a fresh combination ; abc is the same combination as bac or cba and can count only once. Therefore, to find the number of combinations of 6 elements taken 3 at a time, we take the number of permutations, $6 \times 4 \times 5$, and divide by factorial 3! :—

$$\frac{6 \times 4 \times 5}{3!}$$

because factorial 3 is the number of permutations we can arrange with any 3 elements. So

$$_6C_3 = \frac{6 \times (6-1) \times (6-2)}{1 \times 2 \times 3}.$$

We can continue in the same way to write down the number of combinations of any number of elements taken in groups of any size. For example, the number of combinations of 10 elements taken 5 at a time is

$$_{10}C_5 = \frac{10 \times (10-1) \times (10-2) \times (10-3) \times (10-4)}{1 \times 2 \times 3 \times 4 \times 5} = 252.$$

Let us for a moment put on one side the consideration of combination as such and look more closely at this new kind of number, e.g., $_6C_2$ or $_{10}C_5$, which we have introduced. The first thing to notice is that the number on the right-hand side of C can never be greater than the number on the left, for it would be stupid to write $_6C_{10}$ because this would amount to asking us to arrange 6 elements in groups of 10. The next point to be noted is how to arrive at the value of one of these new numbers. The number stands for a division sum ; it is best to put down the denominator (the bottom line) first. Note that there are as many numbers in the top line (the numerator) as in the bottom line. The denominator goes up in steps of 1, beginning with 1, and the numerator descends in steps of 1,

beginning with the left-hand number. Here are some more
examples :—

$$_{17}C_6 = \frac{17 \times 16 \times 15 \times 14 \times 13 \times 12}{1 \times 2 \times 3 \times 4 \times 5 \times 6}$$

$$_8C_7 = \frac{8 \times 7 \times 6 \times 5 \times 4 \times 3 \times 2}{1 \times 2 \times 3 \times 4 \times 5 \times 6 \times 7}$$

$$_{19}C_2 = \frac{19 \times 18}{1 \times 2}.$$

There is another way which is sometimes useful of writing
down these new numbers :—

$$_6C_2 = \frac{6!}{2!(6-2)!} = \frac{6!}{2! \times 4!}.$$

Writing this in full we have

$$\frac{6 \times 5 \times 4 \times 3 \times 2 \times 1}{2 \times 1 \times 4 \times 3 \times 2 \times 1}$$

which is evidently the same thing as

$$\frac{6 \times 5}{2 \times 1}$$

because $4 \times 3 \times 2 \times 1$ occurs both in the top and bottom
line and can be cancelled.

We can use this same method to write down the value of
$_6C_4$ which by the new rule will be equal to

$$\frac{6!}{4!(6-4)!} = \frac{6!}{4! \times 2!}.$$

This is obviously the same as $\dfrac{6!}{2! \times 4!}$, which was the result

given above for $_6C_2$. Therefore $_6C_2 = _6C_4$. How can this be ?
A moment's thought will show that if 6 elements are combined in groups of 2, then, for each group of 2, there corresponds a remainder, a group of 4. So that the number of
combinations of 2 elements is bound to be the same as that
of the remaining 4 elements. In the same way and for the
same reason $_8C_3 = _8C_5$ and $_{12}C_7 = _{12}C_5$. This can be confirmed by calculation if so desired.

When we write down a series of these new numbers, for
example $\qquad _9C_1, \; _9C_2, \; _9C_3, \; \ldots \; _9C_8,$

we now know that the first number is equal to the last, the

second is equal to the last but one, and so on. For the general rule runs thus :—

$$\text{(No. of Elements)} \; C \; \text{(No. in Group)}$$
$$= \text{(No. of Elements)} \; C \; \text{(No. of Remaining Elements)}$$

If we work out the numbers in the above series, we find that they are

9, 36, 84, 126, 126, 84, 36, 9, just as we said they should be.

The peculiar symmetry or regularity of these numbers will come to our notice again when we consider the Binomial Theorem later on.

We can now work out combinations. So far we have talked of the size of the groups and said that $_5C_3$ means that we are to combine 5 elements in groups of 3 as often as possible without repetition. From now onwards it will be convenient to speak of " classes " instead of " groups," so that we shall say —$_5C_3$ is the third-class combination of 5 elements. Similarly, $_8C_7$ is the seventh-class combination of 8 elements, and so on. Let it not be forgotten that if $_5C_3$ is the third-class combination of 5 elements it equals the second-class combination of 5 elements, i.e., $_5C_2$, according to our rule given above.

The possibility of having elements repeated arises in combination just as in permutation. But there is a difference. In permutation we limited the number of similar elements or things, but in combination with repetition we are only going to consider the case in which any one element can be used as often as we like. For example, if we are combining the 5 elements a, b, c, d and e we can allow such combinations as aaa, abb, bbc, dee, ddd, etc. To arrive at the formula which is required for this kind of combination is difficult and wearisome, so we will content ourselves by stating it :—

$$\left(\begin{smallmatrix}\text{No. of Elements} \; + \\ \text{Class number} - 1\end{smallmatrix}\right) \; C \; \text{Class number.}$$

Supposing we had 8 elements from which we were to make up groups of 4 with unlimited repetition. We should write

$$_{(8+4-1)}C_4 = {}_{11}C_4 = \frac{11 \times 10 \times 9 \times 8}{1 \times 2 \times 3 \times 4} = 330,$$

which is naturally a bigger result than if we had no repeated elements in our groups. In that case the number of combinations would only be

$$_8C_4 = \frac{8 \times 7 \times 6 \times 5}{1 \times 2 \times 3 \times 4} = 70.$$

To round off our knowledge of this new symbol of the type $_8C_4$ we should ask ourselves what does $_8C_0$ stand for ? We are being asked to make as many groups as possible of no elements at all out of 8 elements. This looks like a nonsensical request. If we try to apply our rules for calculating combinations we find we are asked to write down the number of groups no times. And yet in mathematics we like to be consistent and. to formulate rules which do always apply. Is it perhaps after all possible to find a value for the symbol $_8C_0$? Yes. We already know that

$$_8C_3 = _8C_{8-3} = _8C_5.$$

If we apply this same rule to $_8C_0$ we get

$$_8C_0 = _8C_{8-0} = _8C_8.$$

Now $_8C_8$ is the number of combinations of 8 elements taken as a group of 8 ; that is to say that the value of $_8C_8$ is 1. That does not mean that " the number of combinations of 8 things taken none at a time is 1 " ; it merely means that for the sake of consistency and convenience in later work we shall be able to say in future that the number of combinations of the zero class for any number of elements is 1.

Now that we have agreed on a value for $_8C_0$ we can write down a complete series of the new symbols—

$$_8C_0, \ _8C_1, \ _8C_2, \ _8C_3, \ _8C_4, \ \cdots \ _8C_8.$$

This is an example of a " Binomial Series."

If we calculate the values of these symbols and add them up we get the sum

$$1 + 8 + 28 + 56 + 70 + 56 + 28 + 8 + 1 = 256.$$

The number 256 is the same as 2^8. The sum of any complete series of this kind is always equal to a power of 2, the index of the power to which 2 is raised being always the same as the number of elements. We can use this result to answer quickly questions of this kind : how many selections can we make from 8 books ? A selection can consist of one book, or of 2 or any number up to 8, but we would hardly count the case where we make no selection at all. The answer is then

$$_8C_1 + _8C_2 + _8C_3 + _8C_4 + _8C_5 + _8C_6 + _8C_7 + _8C_8 = 256 - 1 = 255.$$

So we can make 255 selections.

CHAPTER VIII

OTHER KINDS OF ARRANGEMENT

Up till now in our examination of arrangements we have been considering in the first place cases in which all the elements have to be used and in which it is simply a matter of the re-arrangement of their order. That was permutation. There was also the case in which each element could be used several times (of course at most only as often as the size of the group being arranged would allow). In the second place we considered combination in which only a stated number of the elements might be arranged in groups or classes. We can only say that one group is different from another when it is composed of a different set of elements. We also considered combinations in which each element could be repeated as often as the size of the group allowed. We are going to consider one more case, one in which the ideas of permutation and combination are used together. Suppose we are given the elements a, b, c, d, e and f ; we might for some reason or other decide that pairs like ab or ed should be counted as different from the pairs ba or de. Here we have permutation within a combination ; this is one of the most common forms of arrangement.

The children of the family to whom we have already referred for purposes of illustration decided to go out every day in a boat. Only 5 members of the family could be accommodated each time. At first sight this seemed to present a problem like the arrangement of the card party, but the children argued that justice would not have been done until all members of a boating excursion had had a chance to fill each of the five positions in the boat. So the combination problem of the card party became involved with the permutation problem of the order of seating at the dinner table.

We are now sufficiently practised in mathematics to be able to deal with this problem directly. First we write down the number of combinations of the 12 elements taken in groups of 5,

$$_{12}C_5.$$

Each combination or boating party can now be re-arranged in

different orders to give different seating arrangements. Each combination of 5 will give " factorial 5 " orders or permutations. The total number of possible arrangements is therefore

$$_{12}C_5 \times 5! = \frac{12 \times 11 \times 10 \times 9 \times 8 \times 1 \times 2 \times 3 \times 4 \times 5}{1 \times 2 \times 3 \times 4 \times 5}$$

The " factorial 5 " above and below the line will cancel out, leaving as the answer $12 \times 11 \times 10 \times 9 \times 8 = 95{,}040$. Even if a boating party could be arranged every day it would take 260 years for every member of the family to get his turn! There is a more direct method of calculating the number of arrangements of this kind, a method similar to that which we have already used in the chapter on combination. Here it is. We are given 12 elements. There are 12 ways of taking out groups of 1 from these. To form groups of 2 we select any one of these groups of 1 and join it to an element from the remaining 11. This can be done in 12×11 different ways. To form groups of 3 we select any one of these groups of 2 and join it to an element from the remaining 10. This can be done in $12 \times 11 \times 10$ different ways. And so the process is continued until we have completed the selection of the groups of 5. The total, as we know already, will be $12 \times 11 \times 10 \times 9 \times 8$ $= 95{,}040$.

Another arrangement which we have not yet considered is one which allows unlimited repetition of single elements within the selected group. What do we mean by this ? At present we mean the formation of groups of 1, of 2, of 3, etc., in which the elements within the group are not only permuted but can be repeated as often as the size of the group allows. So groups of 3 made up from the elements 1, 2, 3, 4, 5 and 6, arranged with repetition of the elements, would give results like 111, 123, 321, 211, 335, 616, 422, etc. If we look at these a little closer we may observe that they bear a resemblance to something in an earlier chapter, namely the formation of our system of numbers. This is not only a resemblance ; we have hit upon the basic principle of the formation of numerals. We only need to introduce the additional symbols 0, 7, 8, 9 and to arrange them with repetitions and we have the complete number system in the scale of 10. We must note one reservation however ; the symbol 0 cannot be used in the first place of any number. We shall discuss this matter later.

As we do not yet know how to calculate the number of

permutations when elements can be repeated, we must return
to a closer examination of this kind of arrangement. We will
not concern ourselves for the moment with the system of 10
because it immediately introduces the complication about the
symbol 0. Let us begin with 5 elements, a, b, c, d and e.
Obviously, in spite of the permission to repeat any element,
there can only be 5 groups of 1 :—a, b, c, d, e. Now let us
write down systematically all the possible groups of 2 :—

aa	ba	ca	da	ea
ab	bb	cb	db	eb
ac	bc	cc	dc	ec
ad	bd	cd	dd	ed
ae	be	ce	de	ee

There are 5 × 5 pairs, that is 25 in all. To find the groups
of 3 we can take each pair and combine it with a, with b,
with c, with d and e in turn. This will give us 5 × 5 × 5 =
125 groups of 3. By a similar method we can find out the
number of groups of 4 : 5 × 5 × 5 × 5 = 625. These results
owe nothing to factorial numbers or binomial coefficients.
They are simple powers of 5. The base (as the number is
called which is being raised to a power), is the number of
elements and the power is the size of the group. In the case
which we have just been considering in which 5 elements were
permuted with repetition, we found that the number of

$$\text{groups of } 1 = 5^1 = 5$$
$$\text{,, ,, } 2 = 5^2 = 25$$
$$\text{,, ,, } 3 = 5^3 = 125$$
$$\text{,, ,, } 4 = 5^4 = 625$$

and so on.

Now we will return to our number system. We can see
that any number system based on place values is only a well-
ordered permutation with unlimited repetition of the elements
(the number-symbols) with the one restriction that the
symbol 0 must not come first in any arrangement. Supposing
we overlooked this restriction. What would be the effect ?
It is clear that, when forming groups of 3, for example, from
the 10 elements in our number system, we should be counting
such arrangements as 000, 006, 017. We do not usually write
these numbers in this way. They would normally be 0, 6, 17.
The number of groups of 3 according to our established rule
should be 10^3 or 1,000. This number of groups of 3 includes

all the 10 single-figure numbers not usually written with two 0
symbols in front of them as well as all the double-figure
numbers usually written without a 0 in front.
Let us begin our arrangements with the simplest case and
build up our system of 10 as permutations

Of single-figure numbers . . . . $10^1 = 10$
(Note ; this includes 0.)

Of two-figure numbers . . . . $10^2 = 100$
(Note : this includes all single-figure numbers
prefixed by 0.)

Of three-figure numbers . . . . $10^3 = 1,000.$
(Note : this includes all previous numbers
prefixed by 0.)

The observations made in the notes arise from the unlimited
repetition of 0.

We could write these results down in another way ; we will
decide not to count 0 as a number ; then the total of single-
figure numbers is $10^1 - 1 = 10 - 1 = 9$; the total of two-
figure numbers is $(10^2 - 1) - 9 = 99 - 9 = 90$. (We take
1 away from 10^2 because it represents 00 which we decided
not to count.) The total of three-figure numbers is $(10^3 - 1) -
90 - 9 = 999 - 99 = 900$, and so on. Our formula therefore
has produced the result that there are 9 single-figure numbers,
90 two-figure and 900 three-figure numbers if we disregard 0
as a number. It would not have taken so long to arrive at this
result if we had simply added up the total of the numbers in
each figure-group as we know them, from 1 to 1,000. But it
has not been really a waste of time to use the formula for it
has shown that our system of numbering in the scale of 10 is
simply the permutation of 10 elements. This idea of permuting
a number of elements is true for any system of place values,
no matter what the number of elements ; it is therefore
applicable to a system in the scale of 6 or 2, or in the scale of
any other number.

CHAPTER IX

WE have often found it rather restricting not to be able to generalise more readily when proceeding with our researches. We purposely limited ourselves to the sphere of natural numbers (1, 2, 3, etc.), but the shadow of generalised numbers dogged us. Indeed, at one point we went so far as furtively to use general terms when describing a mathematical operation ; we said :—

$$\text{Dividend} \div \text{divisor} = \text{quotient}$$
$$\text{Divisor} \times \text{quotient} = \text{dividend.}$$

In this formula words stand for any numbers we want to use. We generalised too at another point in our discussion of combination on p. 38. It was rather daring of us as we shall see later, but we will now repent and put our past misdeeds to good purpose. True to our first resolution always to begin with the simplest and most concrete case, let us consider a problem concerned with the area of a room.

We are asked to measure the floor space of a room in order to buy a fitted carpet for it. The room is not very big ; it is 10 feet long and 6 feet wide. It does not take much calculation to appreciate that we could cover the floor with pieces of carpet 1 foot square using either 10 rows of six pieces in a row or 6 rows of ten pieces. We can write down therefore either $10 \times 6 = 60$ or $6 \times 10 = 60$. In conversation we actually do say that the room is " 10 by 6 " or " 6 by 10." We should need then 60 square feet of carpet to cover this room. If we were asked to cover another room and this one was 8 feet by 12 feet, we should multiply 8×12 and buy 96 square feet of carpet—and so on. What do we mean by " and so on," a phrase which keeps cropping up so frequently ? It tells us that the next, and similar, calculations have to be done by applying the same rule ; in other words, we are making a generalisation. In our particular case the rule is : " In order to find the area of a four-sided right-angled figure or rectangle, multiply the length by the breadth." What length by what breadth ? A given length and its corresponding breadth.

We are obviously leading up to a " formula " for the area of a
rectangle and here it is :—

$$A = l \cdot b.$$

In this formula A stands for the area of the rectangle, l for its
length and b for its breadth.   It makes no difference if we
write $A = b \cdot l$ as $b \times l$ gives the same result as $l \times b$.   For
the first time we have extended our methods of calculating
and our operational commands to apply not only to numbers
but to letters which of course represent numbers.

Let us look at some more examples.   We are agreed that
2 apples added to 1 apple make 3 apples.   4 apples and 3
bananas, on the other hand, result in a new quantity, say 7
fruits or a dish of fruit.   In the same way 5 times 3 apples
equals 15 apples and 27 bananas divided by 9 equals 3 bananas.
If we write " a " for apple, " b " for banana, " c " for a fruit
and " d " for a dish of fruit, then we can translate our previous
sentences thus :—

$$1a + 2a = 3a$$
$$4a + 3b = 7c \text{ or } 4a + 3b = d$$
$$5a \times 3 \ = 15a$$
$$27b \div 9 \ = 3b.$$

This certainly looks like some new method of thinking and
calculating but we will not stop now to examine it.   We will
go on to another problem.   What number must be added to
28 in order to make it up to 3 times the number added ?   We
may not immediately know what the required number is.
For the moment it is " the unknown."   Let us call it " x."
So we are to add x to 28.   We know how to write this down :
$x + 28$.   But this sum is to be equal to 3 times the unknown
number.   Therefore we can write

$$x + 28 = 3 \times x \text{ (or } 3x).$$

At this point we should note that we are using the sign $=$ in
a new way.   So far we have used it only as a sign showing that
an operation has been completed.   Now we are saying that
$x + 28$ is to be made the same as $3x$, so that the sign " $=$ "
is now an operational command, demanding "make this
equal to . . ."   But how is this command to be carried out ?
If we search for this number x we find that when we try the
number 14 for x we get $14 + 28 = 3 \times 14$, which is obviously
a correct solution of the problem.

We will pause here only to note that the equals sign is a kind of equalisation machine the rules for the working of which are numerous and which will be dealt with in detail later. For the moment we must consider how it is that the examples of the carpet, the apples and bananas and the unknown number x come to be related.

For the time being we can only make a rather superficial point. We have suddenly left the world in which natural numbers are used and have entered that in which letters are associated with numbers. We are now using the methods we have learnt when calculating with figures for any kind of thing, even for " the unknown." We have already used the letter b in 2 ways. First it was the breadth of a room and secondly it was a banana! But apart from this we can let it be itself. We can say for example that $7 \times 3b = 21b$, whatever we agree to let b stand for. The only restriction is that during any calculation in which " b " is used it is understood that it continues to represent the same thing. $2b + 3b$ only equals $5b$ as long as the bs represent the same thing all the way through the calculation, although we need not know what this thing is. Indeed, when we use " the unknown x " we know only that it is something (usually a number) for which we are seeking. But this fact is in itself sufficient to tell us something about x, to make it no longer unknown. If we say that x is the something for which we are seeking, we have already defined it although we may not know its value as a number without solving a mathematical problem.

In any case it is clear that we have introduced a new notation or system of writing down in which the symbols stand for very nebulous objects or quantities. They certainly cannot be said to represent always one and the same thing as the natural number symbols like 9 or 17 do.

It is just because they have no single meaning that they can be used completely generally. The formula $A = l \cdot b$ gives not only the surface area of our room but of all rectangular rooms ; indeed, of all rectangles. And $3a + 2a = 5a$ is equally true whether the " a "s are apples, pencils or aeroplanes, 5's or 13's, anything, so long as they are similar objects or quantities.

We have reached a new stage in our mathematical progress in which we have been introduced to the idea of generalised numbers of " *Algebra*." The distinction between these and the natural numbers which we already know is the (for the present) vagueness of the generalised numbers.

Before we take a closer look at algebra, let us digress for a moment to mention the origin of algebra and of its name. The Arab mathematician Alchwarizmi, whom we mentioned in the first chapter, was the author of a treatise entitled " L'Al'djebr. . . ." in which he dealt with rules of equalisation which we cannot discuss yet. The title of his book, however, after some ill treatment, became in the course of time our word " Algebra."

It must not be supposed that we in the West took over from the Arabs a perfected system of calculating with general numbers. Development took place over centuries. The first algebra book of which we have a copy was written about 2,000 B.C. in Egypt. The ancient Hindus, Egyptians, and later the Greeks, all had a share in the development. Knowledge of algebra was brought to Western Europe by the Moors after they had invaded and " colonised " part of Spain. During the Renaissance ancient knowledge was eagerly sought, studied, and its ideas developed. Algebra was among the new " arts." Since that time Vieta (born 1540), Descartes (born 1596), and Euler (born 1707) who finally standardised the modern notation, are some of the names famous in the history of algebra.

The reader will have to forgive now a mild digression into philosophical matters but these will not be as difficult as the work we have done, for instance, on number systems other than that based on 10. We want to do more than merely learn the mechanical rules for dealing with algebraic symbols. We want to understand something of the inner structure and basic ideas of algebra.

CHAPTER X

ALGEBRAIC NOTATION

We will return to our problem of carpeting the floor of a room but if we are to make progress in generalisation we must take less interest in the objects themselves and concern ourselves more with their form. In future therefore we shall always refer to a rectangle, this being defined as a four-sided geometrical figure with all its angles right angles. That is the kind of shape for which our formula $A = l \cdot b$ is valid.

Now we have to discover what is meant by the algebraic multiplication (the product) of $l \times b$, or $l \cdot b$, if we want to make general use of our formula. Both l and b can stand for any numbers however small, however great; for example 2×1 or $1{,}994{,}373 \times 284{,}786$. The result of the multiplication will give the surface area, expressed of course in square units although for the present we are not concerned with the nature of these units. Now we ought to find out how far we can apply our algebraic formula. Has it any limitations? To determine this we will begin by fixing the length and letting the breadth first increase in stages and then decrease as far as possible. It will be convenient to do what is usual in this case—to call the longer side of the rectangle the length and the shorter side the breadth; of course if both are the same it does not matter which is which.

Now let us imagine that the breadth of a rectangle is increased in stages or steps, a unit at a time, until the breadth is equal to the length (see Fig. 2).

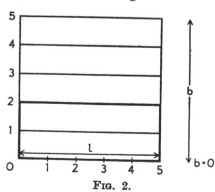

FIG. 2.

The increase cannot be continuous but must go in jumps from one whole number to the next, because so far we have only been dealing with whole numbers. When the breadth is eventually equal to the length, that

48

is when b equals l, our formula becomes

$$A = l \cdot l$$

which we have already learnt to write as l^2 and call either "l to the power of 2" or better "l squared." This case, when the breadth is equal to the length and the rectangle has become a square, is one extreme of all the possible shapes of our rectangle. Let us now examine the other extreme. Let us decrease the breadth in steps of one unit until it is as small as possible, when it equals 0. Then

$$A = l \times 0 = 0.$$

This should not surprise us unduly, because it is clear from our figure that the rectangle has now become a straight line with no breadth at all. The "area" of this line is obviously 0, agreeing with the result of our formula. Our first algebraic formula has proved its worth : now let us try something harder.

We are to limit our rectangle in a second way, namely by making the length 5 units and the breadth 2 units. This fixes the *shape* of the rectangle but its *size* will depend on the units we choose. These could be inches or light-years, leagues or sea-miles, yards or metres. In every case the area of the rectangle by our formula will be $5 \times 2 = 10$ square units of whatever measure we choose (see Fig. 3).

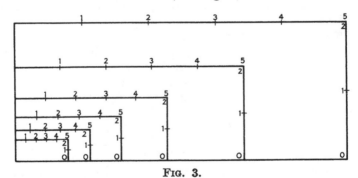

FIG. 3.

Our formula, as we can see from Fig. 3, works for any sort of unit.

There is one more way of testing our formula. Suppose we decrease the length as well as the breadth until both equal 0, the area of such a rectangle can hardly be anything but 0.

What does our formula tell us ? $A = 0 \times 0 = 0$. Our formula still gives the correct answer even in this exceptional case.

Now let us keep the proportion of the length and the breadth of our rectangle at 5 to 2 and at the same time choose smaller and smaller units. The rectangle will eventually become nothing more than a geometrical dot ; both its length and breadth are non-existent, yet they still have the proportion of 5 to 2, and this is very important. It is as though the non-existing length of (5) units of a non-existing rectangle was multiplied by the non-existing breadth of (2) units to produce a non-existing surface area equal to (10) square units. We are asked to imagine a rectangle so small that it is represented by a dot. It has no measurable sides but its sides are still thought to be in the proportion of 5 to 2, and that is its distinguishing feature. This is asking much of the reader's imagination, but he must accept it or he will become involved in other and even greater difficulties.

This step which we are asking the reader to take is one which worried mathematicians from the time of the Greeks. Indeed, the rigidly logical mind of the Greek mathematicians like Euclid recoiled from accepting an idea which required them to take so much on trust. It is only in comparatively modern times that the logical difficulties have been surmounted, but further consideration of these must be deferred to a later chapter.

We ourselves have not recoiled and are ready to proceed with a more mathematical statement of this new idea. The one stipulation concerning our last rectangle was that the length and the breadth had to be in the proportion of 5 to 2 or (in technical terms), " the ratio of the length to the breadth is as 5 is to 2."

That is to say in mathematical shorthand,

$$l : b = 5 : 2.$$

If we reduce both l and b to 0, as we have just done in imagination, then we can write $l : b = 5 : 2$ as

$$0 : 0 = 5 : 2$$

or, if we prefer it,

$$\frac{0}{0} = \frac{5}{2}.$$

If we now assume that any two 0's are the same, then $\frac{0}{0}$ can only be 1 ; for a number divided by itself is always 1. If we accept this, then the equals sign demands also that we accept that $\frac{5}{2}$ is 1—an obvious piece of nonsense. Our only alternative is to imagine that in this case $\frac{0}{0}$ is equal to $\frac{5}{2}$. And so it is that in general $\frac{0}{0}$ is a so-called indeterminate quantity. We only know its value if we know the particular circumstances in which the ratio arises.

There is another way of looking at this ratio $\frac{0}{0}$. It is clear that the ratio 3 : 9, or $\frac{3}{9}$, is equal to the ratio 7 : 21 or $\frac{7}{21}$. If we write it down like this,

$$\frac{3}{9} = \frac{7}{21}$$

there is a well-known rule of " cross-multiplication " that gives

$$3 \times 21 = 7 \times 9 \text{ (both sides being equal to 63).}$$

If we now apply this rule to the ratios $\frac{0}{0} = \frac{5}{2}$ we get

$$0 \times 2 = 0 \times 5,$$

and this is true since both sides of the sign equal 0.

So much for the proportion of 0 to 0 and its threat of higher mathematics ! Let us return with relief to the more solid world of apples and bananas. That 1 apple plus 5 apples make 6 apples should not present a problem. Matters are not quite so simple if we say that 3 apples plus 5 bananas make 8 fruits or even 1 dish of fruit. If we call an apple " a," a banana " b," a single fruit " c " and a dish of fruit " d," then we can write

$$3a + 5b = 8c$$

or

$$3a + 5b = d.$$

Now, if two things are equal to a third thing, then they must be equal to each other. So

$$8c = d,$$

since both $8c$ and d are equal to $3a + 5b$.

This is not a very surprising result, since we began by
defining a dish of fruit (d) as 8 fruits consisting of the sum of
3 apples and 5 bananas. There is not much difficulty in
accepting the idea of a dish of fruit, but when we come to
consider the " c "s, the pieces of fruit, we realise that we have
transformed apples and bananas, two distinct classes of objects,
into one class of objects, namely " pieces of fruit." In mathematics, therefore, we need no longer denote them as " a,"
" b," " c," etc., as long as we are looking upon each of them
purely as one kind of thing, a piece of fruit. We can drop the
letters and write

$$3 + 5 = 8,$$

remembering to read the answer as " 8 *pieces of fruit.*" This
is the same thing as counting a and b as each equal to c and
then performing the calculation $3c + 5c$, which gives us the
result $8c$ simply by adding the " co-efficients."

This provides the opportunity for associating what we
know of algebra with what we already know of numbers.
We know that any number multiplied by 1 remains unchanged.
So our natural numbers can be thought of as 1×1, 2×1,
3×1, 4×1, etc. We can look on our numbers 1, 2, 3, 4 . . .
then as coefficients, not of apples, bananas, nor of letters like
a, b and c, but of " ones." Therefore 5 ones + 17 ones make
22 ones ; 48 ones $\times$ 15 = 720 ones when we think of the
numbers as coefficients of " one." Of course we get the same
result if we use these coefficients as numbers, but this idea of
considering the natural numbers to be coefficients of " one "
makes a bridge for us between the operations of arithmetic
and algebra.

Having constructed this bridge, we are now in a position to
go further into the field of algebra. A question may have
occurred to us—is it possible in algebra to dispense with
natural numbers altogether and to deal only in letters ?
After all, in arithmetic we used only natural numbers and
no letters. In algebra so far we have used a mixture. The
solution of this problem is an attempt to generalise the
co-efficients which hitherto have been numbers. We have
used the expression " coefficient " often before. Let us remind
ourselves of its nature. In the term $5a$, 5 is the coefficient and
" a " the unit in which we are dealing ; similarly in the
expression $5a + 7b + 3c$, 5, 7 and 3 are the coefficients and
a, b and c are the units. These units are general in the sense

that they can stand for anything but they are limited in so far as they must stand for the same things throughout the calculation. For this reason they are sometimes called "*constants*." Convention has ordained that these constants be represented by the letters a, b, c, d . . . in alphabetical order as far as u, with some exceptions. The letters after u, and sometimes r, s and t, are used in another way which will soon concern us. Even some letters at the beginning of the alphabet have to be used discreetly because it has become conventional to use them for special purposes. Since the time of the mathematician Euler, " e " has been used to denote " the base of the natural logarithms " ; since Gauss, " i " has been used to denote " complex " or " imaginary " numbers. The letter " d " has a special use in the Calculus. Any letter of the alphabet is as useful as any other as a constant, but care must be taken not to muddle ourselves by introducing those with special conventional uses at times when a confusion might arise. A more convenient system suggested by Leibniz is used in modern mathematics. Instead of a, b, c, d, etc., we write a_1, a_2, a_3 . . . as many of them as we require for our purpose. (These are read as " a suffix one," " a suffix two," " a suffix three," etc., or if no confusion is likely to arise, as " a one," " a two," etc.)

We have suggested so far that letters can be used only for the constants in algebraic expressions. Can we go further and generalise the coefficients too ? Yes, we certainly can. In the expression

$$a_1 m + a_2 n + a_3 p$$

a_1, a_2 and a_3 would usually be the coefficients and m, n and p the constants. When looking at an expression of this kind it is not always possible to say at a glance which are the constants and which the coefficients, but there is rarely any confusion because we normally begin an algebraic calculation by stating what letters we are going to use and how we intend to use them. Moreover, it is usual, as in the example quoted, to write the coefficient in front of the constant. If we rewrote the example as

$$m a_1 + n a_2 + p a_3$$

it would suggest that m, n and p were being used as coefficients and a_1, a_2 and a_3 as constants. If the example were again rewritten, this time as

$$a_1 a_2 + a_3 m + n p$$

it would create unnecessary difficulties. We should find it hard to remember the purpose for which each letter was being used and the coefficients would not easily be distinguished from the constants.

It should be clear by now that the letters of algebra can be treated in just the same way as the numbers of arithmetic. We can dispense with numbers altogether but we shall in fact find that they crop up in all kinds of ways. For example we know that $a \times a \times a = a^3$; it would not be helpful to replace the index 3 by a letter unless, as very rarely happens, there is some doubt about the scale of our number system. (If we are counting in the scale of 2, $a \times a \times a = a^{11}$.) This does not mean that the index of a number raised to a power can never be generalised ; we can write a^n, which means that a is multiplied by itself n times. We do not propose to linger over these points. It is better to see algebra in action. All that we want to emphasise here is that algebra gives us a means of performing very general calculations in which numerical values are not of the first importance.

CHAPTER XI

THE SIMPLER OPERATIONS OF ALGEBRA

SINCE we already know so much about algebra we ought now to begin to consider its general rules. It is clear above all else that we cannot apply the methods of arithmetic to algebra without some modifications. We cannot construct a true system of place values with letters whose values are not defined nor can we calculate in the same way as we did with numbers. Always bearing our apples, bananas, etc., in mind, let us collect together the fundamental principles which apply to calculation in algebra.

When we write $a + b$ we mean that we are to add together the two different quantities a and b. How are we to achieve this ? If we merely want to do a calculation we can go no further ; $a + b$ remains $a + b$. At most we can say that it might equal $b + a$, because, when adding, the law of commutation (see Chapter III, p. 15) holds good. Similarly $a + b + c$ can only be $a + b + c$ or $b + a + c$ or $c + a + b$. On the other hand, if quantities are alike, we can simplify the expression :—

$$a + a + a + b + b + c = 3a + 2b + c.$$

The coefficient 1 is not usually written so that " a " means $1a$ or $1 \times a$.

We can, if we wish, use the same letter with different suffixes instead of using different letters. The last result could then be written thus :—

$$a_1 + a_1 + a_1 + a_2 + a_2 + a_3 = 3a_1 + 2a_2 + a_3.$$

Now suppose that we have two expressions which we have to add :—

$$
\begin{array}{ll}
 & 3a + 2b + 4c \\
\text{adding} & 2a \quad\;\; + 3c \\
\hline
\text{Total} & 5a + 2b + 7c
\end{array}
$$

As we can see, all that is done is to add the coefficients, but it is only possible to add the coefficients of terms that are alike. What we must not do is to add together those of terms which bear different suffixes or indices ; $a_1 + a_2$ equals nothing

but $a_1 + a_2$ whatever we try to do with it ; similarly, except
for changing the order of the terms, $a^2 + a^3$ can never result
in anything different.

Unlike addition, subtraction is not commutative, that is to
say, $a - b$ is not the same as $b - a$. It can only be rewritten
$- b + a$. Like addition, however, subtraction is carried out
by dealing with the coefficients of terms which are alike :—

$$
\begin{array}{l}
\quad\quad\quad\quad 3a + 2b + 4c \\
\text{subtracting} \quad 2a \quad\quad\;\; + 3c \\
\hline
\text{Total} \quad\quad\quad a + 2b + c
\end{array}
$$

Note that the result of subtracting $2a$ from $3a$ is $1a$, not
just 1.

In the following example we meet a new problem ; we are
asked to subtract $a + 6b$ from $3a + 4b$. Setting this out,

$$
\begin{array}{l}
\quad\quad\quad\quad 3a + 4b \\
\text{subtracting} \quad a + 6b \\
\hline
\quad\quad\quad\quad 2a + \;?
\end{array}
$$

What can we do with $4b - 6b$? How can we take 6 pears
away from 4 pears ? It is really quite simple. Supposing a
shopkeeper were making up his accounts. He has cheques for
$100 and he owes $120 ; that is to say, he owns $100 and
owes $120. He enters this. So he is $20 down ; he owns
"minus $20" worth of cheques. It may not be customary in
business circles to put it quite in this way, but in mathematics
it certainly is. Our sum therefore comes to $2a - 2b$, because
when $6b$ is taken from $4b$ there is a difference of $2b$ still owing.
If we were dealing only with $4b - 6b$ the answer would be
written as $- 2b$.

So we have hit upon the conception of negative numbers
which can be either figures such as $- 2$, $- 4$, or generalised
algebraic numbers such as $- a$, $- bc$, etc. We can visualise
them if we wish as sections of a straight line (see Fig. 4).

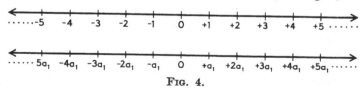

FIG. 4.

Zero is not in itself a number though it is often treated as if it were ; it separates the positive from the negative numbers. A positive or " plus " number is not usually given its plus sign, but the negative or minus sign may not be omitted from a negative number.

We now need an introduction to the use of brackets before any further progress in algebra can be made ; they were used without comment in the chapter on binomial coefficients. It is time to look at them in greater detail.

A pair of brackets signifies that everything inside them belongs together and must be treated as a whole. If we take out the contents of the bracket, the bracket is finished with and we pay no more attention to it. We deal with the bracket and its contents as though it were one independent term.

Consider this expression :—

$$10,000 - (5,020 + 23 - 448).$$

What does it equal ? First of all, deal only with the figures inside the brackets ; work out that sum and arrive at the number 4,595. We can now write the expression again, this time without brackets :—

$$10,000 - 4,595 = 5,405.$$

If we had ignored the brackets and worked the sum out as though it were written

$$10,000 - 5,020 + 23 - 488$$

the answer would have been 4,555 and it would have been wrong.

Let us try another example with brackets :—

$$15,375 - 320 + (8,220 - 26 + 400) = 15,055 + 8,594 = 23,649.$$

Although the rules forbid it, supposing we wrote out this example, omitting the brackets,

$$15,375 - 320 + 8,220 - 26 + 400$$

we should nevertheless arrive at the correct answer, namely 23,649. How does this come about ? May we then ignore brackets after all ? Yes, but only when the bracket is immediately preceded by a plus sign. Should it be preceded by a minus sign, we may not omit the brackets, at least not until we have finished calculating the sum within the brackets or have made certain alterations inside them. To explain this we shall have to take a closer look at negative numbers and their properties. For this purpose, all numbers shall be given their signs, plus or minus, as the case may be. We will put brackets round the number and its sign, *e.g.*, $(+5)$, (-7), to show that they belong together and that the sign is not intended to convey instructions about operations to be undertaken. $5 + 7$ will now be written $(+5) + (+7)$ which equals $(+12)$. Similarly $(+12) - (+7) = (+5)$.

Now look at Fig. 5 :—

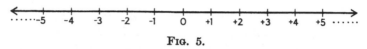

FIG. 5.

We will try to explain the way in which addition and subtraction work on this line before we grapple with negative numbers.

The process of adding along the line can take place in two directions. If we are adding together numbers which are on the right-hand side of zero, that is to say positive numbers only, we must keep going in the same direction, moving towards the right as long as we continue to add, *e.g.*, $(+1) + (+2) = (+3)$, etc. On the other side of zero, as though the picture were seen in a mirror, we move along towards the left when we add the negative numbers, so that $(-1) + (-2) = (-3)$. It is as though we were adding up our debts.

When we subtract the operation takes place in the same way but we move along the line *towards* zero for both positive and negative numbers ; according as we are dealing with $(+)$ or $(-)$ numbers, we move in towards zero from the right or the left. Here are two examples :—

$$(1) \quad . \quad . \quad (+4) - (+3) = (+1)$$
$$(2) \quad . \quad . \quad (-4) - (-3) = (-1)$$

In this second example we have subtracted one debt from another and are left with a smaller debt : we are moving towards solvency, which is another way of describing zero. If we are to be rid of our debts altogether and be left with some cash in hand, circumstances will have to alter. We shall have to pass zero and attain the positive side of the line. Suppose we had to add $(+3)$ and (-2). We have cash to the value of $300 and debts of $200. We pay off the debts and have $100 left, or

$$(+3) + (-2) = (+1).$$

It is as though we cut the line of numbers at the spot where 0 is and then laid one part parallel to the other, making the minus quantities in the upper half correspond to the plus of the lower half.

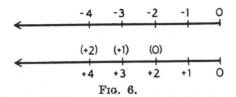

FIG. 6.

When debts and cash in hand have cancelled each other out, we note whether we are left with any of one or of the other. We have established a new spot on the line for 0 ; 0 is now at the point where cash in hand and debts are equal, so we must write down the new 0 there in brackets and remember the line. In our particular example where we are adding $+ 3$ and $- 2$, the new 0 in brackets will come at $+ 2$ and the surplus of cash in hand will be found on the " plus " part of the line. We then write down the fresh numbers in brackets above the line starting from the new 0. The bottom number ($+ 3$ in this case) shows how far we have to go ; the number above it in brackets gives us the answer ; it shows us that

$$(+ 3) + (- 2) = (+ 1)$$

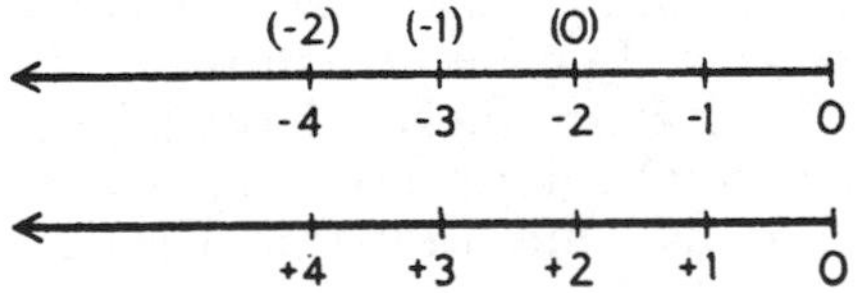

FIG. 7.

In the same way we can show by Fig. 7 that

$$(- 3) + (+ 2) = (- 1).$$

When we are dealing with subtractions which cross the frontier of the minus area we must be particularly careful. Take the expression $(+ 2) - (- 1)$.

We are given \$200, $(+ 2)$. We are told also to write off or subtract debts of \$100, $(- 1)$. We have not only been given

$200 but we have had $100 of debts cancelled as well. We are therefore $300 or $(+3)$ to the good. In short,

$$(+2) - (-1) = (+3).$$

Our next step is to collect all the types of expressions with which we have been dealing (and some others too), write them down one under the other and try to draw some general conclusions about them.

$$(+1) + (+2) = (+3) \qquad (+3) + (-2) = (+1)$$
$$(-3) + (-1) = (-4) \qquad (-3) + (+2) = (-1)$$
$$(+4) - (+3) = (+1) \qquad (+2) - (-1) = (+3)$$
$$(-4) - (-3) = (-1) \qquad (-2) - (+1) = (-3)$$

Look at this table carefully. It will soon be seen that it is possible to avoid the use of brackets and arrive at the correct solution of the expression.

$$+1 + 2 = +3 \qquad\qquad +3 - 2 = +1$$
$$-3 - 1 = -4 \qquad\qquad -3 + 2 = -1$$
$$+4 - 3 = +1 \qquad\qquad +2 + 1 = +3$$
$$-4 + 3 = -1 \qquad\qquad -2 - 1 = -3$$

How has this happened ? The sign which precedes the number and tells whether it is positive or negative, becomes involved with the sign which tells what operation to perform, adding or subtracting. For example,

$$(-4) - (-3) = (-1) \text{ has become } -4 + 3 = -1.$$

The minus sign denoting " subtract " has merged with the minus sign denoting " negative number." The operational command and the descriptive sign have joined together.

Here are the general rules concerning this merging :—

If both operational and descriptive signs are alike, the result is plus.

If they are different, the result is minus. These rules hold good not only for one number within brackets but also for several.

$$20 - (3 - 7) = 20 - 3 + 7 = 24.$$

If we work out the contents of the bracket the result is

$$20 - (-4) = 20 + 4 = 24.$$

That is why we can omit the brackets, as we mentioned earlier, if there is a plus sign in front of the bracket. For in the bracket a plus sign will remain plus because both signs are alike, and a minus sign will remain minus because the two signs are different.

Now let us move on to deal with brackets within brackets. We have already used them once without further explanation. Here is an example :—

$$17 - \{8 - (6 + 4)\}.$$

To find the answer to this, begin with the innermost bracket and simplify that first.

Thus
$$17 - \{8 - (6 + 4)\}$$
$$= 17 - \{8 - 10\}.$$

Then
$$17 - \{8 - 10\} = 17 - (-2)$$
$$17 - (-2) = 17 + 2$$

and finally
$$17 + 2 = 19.$$

It is high time that we returned to the study of algebra. We were in a somewhat bewildered state when we left it in order to discuss subtraction. This operation will not in future cause us any anxiety now that we understand that if we have 2 apples and give away 2, none remain ; if we have 6 and give away 4, 2 remain ; and if we have 3 apples and have to give 7 away, we owe 4 apples.

Having generalised them, we can go on to deal with brackets without laboriously working through the intervening stages ; and we can also introduce different quantities into our next examples.

(1)
$$x^2 + \{y^2 - (x^2 + y^2)\}$$
$$= x^2 + \{y^2 - x^2 - y^2\}$$
$$= x^2 + (-x^2)$$
$$= x^2 - x^2 = 0$$

(2)
$$3a - \{2b - (a - b)\}$$
$$= 3a - \{2b - a + b\}$$
$$= 3a - \{3b - a\}$$
$$= 3a - 3b + a$$
$$= 4a - 3b.$$

As you can see, it is usually only possible fully to work out the contents of brackets in algebra when the quantities are alike.

Before we leave algebraic addition and subtraction it must be pointed out that really these two operations are interchangeable. This leads us to the idea of the " algebraic sum," which consists of adding quantities with plus and minus signs together and abolishing subtraction. $5 - 3 - 2 + 4$ can be rewritten as $+ (+ 5) + (- 3) + (- 2) + (+ 4)$ and therefore be regarded purely as an exercise in addition.

Having made fair progress in generalising, we can go on to

consider something new, namely the distributive principle. We are agreed that

$$(7 + 4 - 3) = 8$$

so that

$$5(7 + 4 - 3) = 5 \times 8 = 40.$$

But we could also get the answer if instead of working out the bracket first, we wrote

$$5(7 + 4 - 3) = 5 \times 7 + 5 \times 4 - 5 \times 3$$
$$= 35 \quad + 20 \quad - 15$$
$$= 40$$

We do not usually do this multiplying out in arithmetic because it is simpler to deal with the contents of the bracket first. In algebra, however, we may not be able to simplify the contents of the bracket in a similar situation, so that when we write the same kind of expression in general terms we have

$$a(b + c - d) = ab + ac - ad.$$

Whatever quantity stands *immediately* in front of the outside bracket multiplies each individual term inside. In order to make quite sure that we understand this, we will try to illustrate what we mean by this distributive principle by drawing a picture of it.

$a(b + c + d)$ can be illustrated thus :—

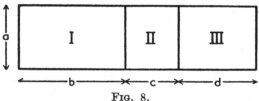

FIG. 8.

Rectangles I, II and III each have the same breadth, a ; the lengths b, c and d are different. The area of rectangle I is obtained by multiplying side a by side b ; similarly the areas of rectangles II and III are $a \times c$ and $a \times d$ respectively. If we want to find out what the area of the whole picture is, we can do it in two ways. We can either add up the areas of the three separate rectangles or we can add the lengths of sides b, c and d and multiply the result by a. In other words, what we can do is to write either

$$a \times b + a \times c + a \times d$$

or

$$a(b + c + d)$$

so that the " a " outside the bracket multiplies all the quantities inside.

If the reader is so minded he can verify for himself the action of the distributive principle when minus signs occur inside the bracket. Let him cut out the positive rectangles and place them side by side. Then let him cut out the negative rectangles and lay them on top of the positive ones, thereby blotting them out, or in other words, subtracting them. He will find that he will get the same result if he multiplies the base line of the large complete rectangle by the side a. He calculates the base line by adding and subtracting its sections as required.

Now we can proceed to apply the distributive principle to expressions which contain not only one term " a," but more. We can still draw our rectangles and if we calculate the areas

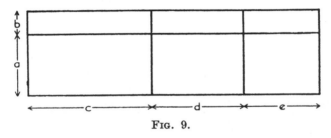

Fig. 9.

of these, we find that the areas of the six rectangles equal $ac + ad + ae + bc + bd + be$. If we now look at Fig. 9 as one large rectangle, however, we have

$$(a + b) \times (c + d + e)$$

which when multiplied out becomes

$$ac + ad + ae + bc + bd + be$$

as before.

This is where we come upon the fundamental rule of algebraic multiplication ; to multiply two expressions together you multiply *every* term of one expression by *each* term of the other expression in turn. Remember to pay due regard to the signs ; two similar signs $= +$, dissimilar $= -$.

Here is an example :—

$$(3a + 2b)(4f + g) = 12af + 3ag + 8bf + 2bg.$$

This cannot be simplified any further. We multiplied together the actual numbers in each term so that the 4 of $4f$ was multiplied by the 3 of $3a$ to make 12. When multiplying concrete numbers, they always unite to produce a new number. But $a \times f$ cannot do this, for the original terms can still be seen in the answer af. Algebraic quantities can sometimes unite to produce a new shape when multiplied. Take for example the expression

$$(2a + 3b) \times (5a + 7b)$$

This becomes, when multiplied out,

$$10a \cdot a + 15a \cdot b + 14a \cdot b + 21b \cdot b.$$

[The dot, $a \cdot a$, between the terms represents a multiplication sign ; it is often neater to use.]

We already know a shorter way of writing $a \cdot a$ and $b \cdot b$, namely a^2 and b^2, so that the expression now becomes

$$10a^2 + 15ab + 14ab + 21b^2.$$

Here there is a possibility of further simplification for $15ab + 14ab$ equals $29ab$. The final result therefore is

$$10a^2 + 29ab + 21b^2.$$

So far we have been dealing only with the multiplication of terms with plus signs. The next step is to consider how to multiply if some minus signs occur. When we multiply $a \times b$ we are really multiplying $(+ a) \times (+ b)$. What happens when we multiply $(+ a) \times (- b)$?

In the paragraphs on addition we agreed that $+(+a) = +a$, that both $+(-a)$ and $-(+a) = -a$ and that $-(-a) = +a$. When we multiply $(+ a) \times (+ b)$ we get $+ (+ ab)$, more usually written ab. When we multiply $(- a) \times (+ b)$ we get $-(+ab)$ which equals $-ab$. Finally, multiplying $(-a) \times (-b)$ we get $-(- ab)$ which is $+ ab$. Summarising these results we have

$$(+ a) \times (+ b) = + ab$$
$$(+ a) \times (- b) = - ab$$
$$(- a) \times (+ b) = - ab$$
$$(- a) \times (- b) = + ab.$$

When two expressions, $(a - b)$ and $(c - d)$, are to be multiplied, we can write them again as $a + (- b)$ and $c + (- d)$ and then use the distributive principle as before, so that

$$\{a + (- b)\} \{c + (- d)\}$$
$$= ac + (- b)c + a(- d) + (- b)(- d)$$
$$= ac - bc - ad + bd.$$

Here is another example :—

$$(-a)(-b)(-c) = -(-ab)(-c)$$
$$= +ab(-c)$$
$$= -abc.$$

This leads us to the rule that, when a number of terms are multiplied together, the sign of the result is plus when the number of minus signs is even, and minus when the number of minus signs is odd. A further example should make this clear :—

(1) $(-a)(-b)(-c)(+d) = -abcd.$
    Odd number of minus signs.

(2) $(-a)(-b)(-c)(-d) = +abcd.$
    Even number of minus signs.

After the multiplication has been completed it may be possible to simplify the result. For instance

$$(a-b)(2a+3b) = 2a^2 - 2ab + 3ab - 3b^2$$
$$= 2a^2 + ab - 3b^2.$$

We can only add and subtract terms that are exactly alike. Great care must be exercised not only in the simplification but also in the actual writing of the expressions so that no confusion can arise. We can add $7a^2$ and $4a^2$ or subtract $3abc^2$ from $5abc^2$ but we cannot simplify in the same manner $5abc^2 - 3ab^2c$ because the terms abc^2 and ab^2c are not alike.

The next step to be considered is the use of indices. We already know what is meant by a power and by an index. The index (or exponent) gives us the command to use the number (or base) to which it is attached as a factor as many times as it indicates. Thus $a^2 = a \times a$, $a^5 = a \times a \times a \times a \times a$.

How are powers multiplied together ? Supposing we wanted to multiply $a^2 \times a^5$, we could write the expression

$$(a \times a) \times (a \times a \times a \times a \times a).$$

As we are concerned only with multiplication here, we can write this as

$$a \times a \times a \times a \times a \times a \times a = a^7.$$

As long as the base remains the same, multiplication of

quantities raised to a power is effected by simply *adding together* the indices so that $a^4 \times a^5 = a^9$ and $b^{10} \times b^6 = b^{16}$. We can generalise this as follows :—

$$a^n \times a^m = a^{n+m}.$$

What happens should we wish to raise a power itself to a further power ? Take the term a^3 and raise it to the power of 4, $(a^3)^4$. What do we really mean by this ? We can write it

$$a^3 \times a^3 \times a^3 \times a^3 = a^{3+3+3+3} = a^{12}.$$

It is true therefore that when we raise a power to a power, the two indices are *multiplied together*. As a general statement of this rule we can say that $(a^n)^m = a^{nm}$.

To complete our section on indices let us repeat that any number with the index 0 equals 1, so that when dealing with general terms $a^0 = 1$; and that a^1 is more normally written " a " without the index 1. The letter a can assume any value from 1 to any large number we care to choose. Further, $a^0 . a^n = a^{0+n} = a^n$, because $1 . a^n = a^n$, and $a^n . a$ is obviously $a^n . a^1 = a^{n+1}$.

Division is the last section of our present enquiry into the workings of algebra. How can general expressions be divided ? What happens when we want to divide a by b ? Nothing more than the statement that a is divided by b ; we can write it $a \div b$ or $\frac{a}{b}$, but it must be realised that the quantities cannot be rearranged as in multiplication, where $a \times b$ is the same thing as $b \times a$. When we wish to work out an expression like $a \div b$ we can do no more to simplify it until we can " substitute." That is to say, we must know the numbers which the algebraic terms represent and then substitute the former for the latter. For example, suppose that $a = 12$ and $b = 3$, then $a \div b = 12 \div 3 = 4$. What happens when $3a$ is to be divided by a ? We are really asking ourselves how many times 1 apple is contained in a collection made up of 3 apples. Of course the answer is 3, so that our expression is $3a \div a = 3$. We can write this more generally thus :—

$$n \times a \div a = n.$$

Further examples show that

$$15a \div 3a = 5$$
and
$$ba \div a = b.$$

Because any number divided by itself always gives the answer 1, $a \div a = 1$.

In a little more complicated form, $25abcd \div 5ac = 5bd$. The accuracy of this result can be checked by multiplying the quotient and the divisor,

$$5bd \times 5ac = 25abcd.$$

If we use these examples as patterns to follow, we shall have no difficulty with division in algebra.

The behaviour of indices in division is, as would be expected, related to what happens in multiplication. When we were multiplying similar terms with indices, we found that the indices were simply added together. In the same way, when we divide, they are subtracted ; $a^5 \div a^2 = a^3$. We must take care to subtract the index of the divisor from that of the term to be divided. As an illustration, two further examples of division with indices are given :—

(1) $12a^4b^2 \div 3a^2b = 4a^{4-2} \times b^{2-1} = 4a^2b$

(2) $6x^3y \div 2x^3 = 3x^{3-3}y^1 = 3x^0y^1 = 3y$ since $x^0 = 1$.

We shall need all our newly acquired knowledge of algebra to solve the last problem of this chapter, algebraic long division. It is convenient at the outset to arrange the terms of both dividend and divisor in either ascending or descending powers. The point of the rearrangement is that the order of both should match ; so that if we wish to divide $a^3 + 11a - 6a^2 - 6$ by $a - 2$, we shall first rearrange the dividend in descending powers to become $a^3 - 6a^2 + 11a - b$, since the divisor is already in descending order. The two parts of the expression are now easier to deal with. Here is a further example of rearrangement, preparatory to working out an algebraic long division sum :—

$$8x^3 - 2a^2x - 12ax^2 + 3a^3 \div (x - a)$$

rearranged to match the divisor in descending powers of x we have

$$8x^3 - 12ax^2 - 2a^2x + 3a^3 \div (x - a)$$

In order to show how the actual calculation of algebraic long division is carried out, we will now consider an example in detail :—

$$8x^3 + 3a^3 - 2a^2x - 12ax^2 \div (2x - a).$$

Rearranged and set out :—

$$2x - a) \; 8x^3 - 12ax^2 - 2a^2x + 3a^3 \; (4x^2 - 4ax - 3a^2$$

$(\times 4x^2) \quad 8x^3 - 4ax^2 \qquad \text{(Subtract)}$

$$\overline{}$$

$ 0 - 8ax^2 - 2a^2x$

$(\times - 4ax) \quad - 8ax^2 + 4a^2x \qquad \text{(Subtract)}$

$$\overline{}$$

$ 0 - 6a^2x + 3a^3$

$(\times - 3a^2) \qquad\qquad - 6a^2x + 3a^3 \qquad \text{(Subtract)}$

$$\overline{}$$

$$ 0 \qquad 0$$

The quotient is therefore $4x^2 - 4ax - 3a^2$.

What steps have been taken to arrive at this answer ?

(1) The dividend was rearranged in descending powers of x to match the divisor.

(2) The first term of the divisor was divided into the first term of the dividend and the result was put down as the first term of the final answer.

(3) The divisor was multiplied by this term and the result of the multiplication was subtracted from the first two terms of the dividend.

(4) The third term of the dividend was brought down and written alongside the remainder from the subtraction.

(5) The same procedure as in (2) was carried out using the remainder from the previous subtraction together with the third term of the dividend as the expression to be divided.

(6) The process was continued as in items (2) to (5) until all the terms of the dividend have been brought down and dealt with.

(7) There may of course be a remainder at the end of the algebraic long division, just as there often is when dividing with numbers.

Be careful with the signs. It is important to realise that you are dividing or multiplying by $+ a$ or $- a$, not just by "a" alone : also do not forget to change the signs when subtracting.

Let us examine a further example :—

$$x + a)x^3 - a^3(x^2 - ax + a^2$$
$$x^3 + ax^2$$

$$0 - ax^2 \quad . \quad . \quad . \quad . \quad . \quad . \quad . \quad . \quad (1)$$
$$- ax^2 - a^2x$$

$$0 + a^2x - a^3$$
$$+ a^2x + a^3$$

$$0 - 2a^3 \text{ Remainder.}$$

Be it noted that here we are left with a remainder. It can be seen from this example that at line (1) it is impossible to bring down the next term of the dividend and subtract from it. Were we to attempt it we should find that we had to subtract $- a^2x$ from $- a^3$ which are two dissimilar terms. This is impossible, therefore we leave the next term of the dividend alone until such time as it happens to fit into our subtraction calculation.

Here we will pause a moment. Having mastered addition, subtraction, multiplication and division, and the use of indices, our present elementary knowledge of algebra is sufficient to carry us on to higher reaches of mathematics.

CHAPTER XII

COMMON FRACTIONS

THIS chapter is concerned not so much with something new as with a different way of writing something we already know—the algebraic way of dealing with division.

We can look upon a fraction either as being an uncompleted division sum or one which has reached the limits of possible expression in whole numbers.

It will make a change to begin with a generalised fraction and write in algebra the fraction $\frac{a}{b}$ which can be read as "a over b" or "a divided by b." The upper part of the fraction is called the "*numerator*" and the lower part is called the "*denominator*." When the numerator is 1 so that we can write fractions in general terms like $\frac{1}{a}, \frac{1}{b}, \frac{1}{c}$, etc., we call such fractions "*unit-fractions*." If the numerator is smaller than the denominator the fraction is called a "proper fraction" and we can write this in general terms as $\frac{a}{b}$ where a is $<b$.

This is the first time we have met the symbols of inequality ($>$, $<$). They are used always with the point towards the smaller quantity so that $a < b$ is read as "a is less than b," and $p > q$ is "p is greater than q."

To return to fractions, $\frac{a}{b}$ is a "*proper fraction*" if $a < b$ and is called an "*improper fraction*" if $a > b$. Here are some examples with numbers : $\frac{1}{3}$ is both a unit-fraction and a proper fraction ; $\frac{2}{3}$ is a proper fraction ; $\frac{7}{2}$ is an improper fraction which could be written partly in whole numbers and partly as a fraction, $3\frac{1}{2}$.

In the mathematics of the ancient Egyptians and of the Greeks, unit-fractions played a particularly important part. The Egyptians for example preferred to write the fraction $\frac{7}{29}$ as the sum of the following unit-fractions : $\frac{1}{6}+\frac{1}{24}+\frac{1}{58}+\frac{1}{87}+\frac{1}{232}$.

A similar method of expressing fractions was used by the Greeks. Nowadays we do not find it necessary to put all our fractions in this form because we have learnt how to deal with all kinds of fractions. Only in the integral calculus is it occasionally convenient to express fractions as " partial fractions," which is a rather similar use of unit-fractions.

Fractions present to us for the first time a new way of looking at numbers. $\frac{1}{3}$ is a third part of one ; $\frac{5}{7}$ is a seventh part of five, and so on. This is something more than division as we have known it up till now. It is as though our whole numbers were being broken into pieces. When we are asked to divide 12 by 3 we are dealing with whole numbers ; all we have to do is to discover how many groups of 3 can be made up from 12. With fractions the problem is different. We are being asked how many bits of a whole number we can squeeze in between one whole number and the next. The fractions $\frac{1}{3}$, $\frac{2}{3}$, $\frac{3}{4}$, $\frac{5}{7}$, $\frac{29}{38}$ all lie between 0 and 1. We can squeeze as many of these fractions between 0 and 1 as we please, provided that we always keep the numerator smaller than the denominator. All proper fractions lie thus between 0 and 1 and, for a given numerator, the greater the denominator the smaller is the value of the fraction. For example, $\frac{2}{25}$ is obviously greater than $\frac{2}{125}$. If the numerator should equal the denominator, this is merely another way of writing 1, for $\dfrac{a}{a}$ is $a \div a$ which is 1. If the numerator is 0 the fraction is then $\frac{0}{b}$; if we divide nothing into bits we still get nothing, so the answer is bound to be 0. If on the other hand the denominator is 0 we are confronted with a much more difficult problem. We are asked to divide something by nothing. So $a \div 0 = ?$ Another way of finding an answer to this would be to ask ourselves what number when multiplied by 0 is equal to a. Obviously there is no such number for we know that any number, however big, when multiplied by 0 gives the answer 0. We can tackle this problem in the following way : let us divide 5 by 100. We get 5 hundredths. If we divide 5 in turn by 47, 21, 7, we get answers which grow larger as the divisor grows smaller. Suppose we continue this process using divisors less than 5. If we divide by 3 we get $\frac{5}{3}$ or $1\frac{2}{3}$. If we divide by 2 we get $\frac{5}{2}$ or $2\frac{1}{2}$; dividing by 1 we get 5. If we continue our divisions now using fractions that lie between 1 and 0, $e.g.$, $\frac{5}{20}$, $\frac{5}{365}$, etc., then as the fraction grows smaller the result of the division rapidly increases. For $\frac{5}{20}$ is already as large as

100, $5 \div \frac{1}{365}$ is 1825, $5 \div \frac{1}{100000}$ is 500,000. Since 0 is smaller than the smallest possible fraction, it must be smaller than the unit fraction which has the largest possible denominator. The fraction $\dfrac{5}{\underset{\text{(a very large number)}}{1}}$ gives 5 × (a very large number), that is to say a fivefold very large number. However large the number may be, the unit fraction with this number as denominator will always be greater than 0.

Let us try to write down this conclusion in a more mathematical form. What we are saying is that a fraction, $\dfrac{5}{b}$ for example, gets larger and larger as b gets smaller and smaller. In fact, by making b small enough we can make that fraction larger than any imaginable number. Here is something which our present stock of mathematical shorthand is incapable of expressing. We shall have to introduce some new symbols and phrases. The mathematical way of saying that a number gets larger and larger without limit, is that " *it tends to infinity.*" This is written $\to \infty$, the arrow standing for " tends to " and the figure eight on its side standing for the word " infinity." Similarly, if a number gradually becomes smaller and smaller it is said to " tend to zero (0)," which is written $\to 0$. There is one other symbol which we had better learn at this stage. It is Lt, short for " Limit." This means that we are being warned that there is a steadily increasing or decreasing process to be undertaken in the calculation. So we can summarise all that we have done in the previous paragraph by writing our former words in this new shorthand :—

$$\underset{b \to 0}{Lt}\, \frac{5}{b} \to \infty.$$

Translating this as it stands we might say :—
" Lt means expect a gradual increase or decrease."
" Of what ? "
" Of b decreasing towards zero, $b \to 0$."
" What next ? "
" As this process happens, the value of $\dfrac{5}{b}$ increases towards infinity, or if we prefer it, increases without limit, $\to \infty$."

Translating our expression more concisely we say that as b decreases to zero, the fraction $\dfrac{5}{b}$ tends to infinity.

This discussion does not appear to have much to do with our chapter heading " Common Fractions." It is much more advanced. Yet we would not have it otherwise. We prefer to meet our difficulties as they arise. Too often the student of elementary mathematics is told that a subject is too difficult for him and the difficulties are wrapped up in mystery ; a little courage and trust in the powers of understanding of the reader would make mathematics much more interesting for him. So we have not hesitated to leap forward into more advanced spheres in order to fill out our knowledge of fractions. Let us now continue our examination of them.

Before we turn to calculations with fractions we will investigate more closely what we are doing when we write " the ratio of a to b " or $a : b$, as $\frac{a}{b}$. Let us take as an example a table-top which when measured is found to be 6 feet long and 2 feet wide. Anybody can see that the length is to the width as 6 is to 2. And since $6 : 2 = 3 : 1$ the table is three times as long as it is wide. What we are doing here is to take one of the two measurements as the unit of measurement and then to compare the other with it. We could equally well have said that the ratio of the width to the length is as $2 : 6$ or $\frac{2}{6}$ or $\frac{1}{3}$; that is, that the width is $\frac{1}{3}$ of the length. In either case we are finding out how often the unit we chose to measure in goes into the other measurement. In the first case the width was the unit and the length was 3 units. In the second case the length was the unit and the width was $\frac{1}{3}$ of the unit. Whatever the units chosen the proportion of the length to the width remains the same.

This relationship between the measurements of the length and width of a table expressed as $6 : 2$ or $3 : 1$ or $\frac{3}{1}$ is a ratio or proportion.

Now we must consider methods of calculating with fractions. First, addition and subtraction ; fractions, as we have already established, are new numbers squeezed in between the whole numbers ; the proper fractions lie between 0 and 1 according to their value the improper fractions elsewhere in the series of numbers. We can of course place plus and minus signs in front of fractions, as we have done in front of whole numbers.

Looking at fractions in a general way without worrying whether they are proper or improper, we can consider them as new numbers. The important characteristic of a fraction is its denominator. It gives its name to the fraction. The symbol $\frac{3}{7}$ is called three " *sevenths* " ; so we can think of 3 as being a coefficient and of the fraction as $3(\frac{1}{7})$. These simple ideas show us the method of dealing with fraction sums. There is one basic principle. Only those fractions which have the same name (or denominator) can be added or subtracted. They can be combined over a common denominator in the following way :—

$$\tfrac{2}{7} + \tfrac{6}{7} - \tfrac{3}{7} = \tfrac{5}{7}.$$

If the fractions should have different denominators then in some way we must find a means of putting them over a common (the same) denominator. We shall assume that the method of achieving this using the " least common multiple " is well known and that an example will suffice :—

$$\tfrac{1}{3} + \tfrac{2}{5} - \tfrac{3}{8} = ?$$

The common denominator here is $3 \times 5 \times 8 = 120$. In order accurately to write each fraction with this common denominator the numerator of each must be changed so that the fraction keeps its original value. $\frac{1}{3}$ written in hundred and twentieths will be $\frac{40}{120}$ because we get $\frac{1}{3}$ again if we simplify this fraction. It would be more practical to determine by what number the original numerator should be multiplied in order to keep the original value of the fraction. Since

$$120 = 3 \times 5 \times 8, \ \tfrac{1}{3} = \frac{1 \times 5 \times 8}{3 \times 5 \times 8}.$$

The full calculation would therefore be

$$\frac{1 \times 5 \times 8}{3 \times 5 \times 8} + \frac{3 \times 2 \times 8}{3 \times 5 \times 8} - \frac{3 \times 5 \times 3}{3 \times 5 \times 8} = \frac{40 + 48 - 45}{120} = \frac{43}{120}.$$

Written in general numbers, an addition and subtraction of fractions would look like this :—

$$\frac{2a}{3b} + \frac{3ac}{2b^2} - \frac{c}{6a}$$

$$= \frac{2a \cdot 2ab + 3ac \cdot 3a - c \cdot b^2}{2 \cdot 3a \cdot b^2}$$

$$= \frac{4a^2b + 9a^2c - b^2c}{6ab^2}.$$

This example should illustrate sufficiently the method of adding and subtracting algebraic fractions and the way in which they are put over a common denominator. If the denominators are all different the common denominator is the product of all of them ; in such a case each numerator has to be multiplied by all the denominators except its own. By this means all the fractions take on the same name and remain equal to their original value.

Remembering that we can say a proper fraction consists of a unit fraction with the same denominator, together with a coefficient (which is the numerator), we can state a simple rule for multiplication.

A fraction is multiplied by multiplying its coefficient. Three times one-seventh is $3(\frac{1}{7}) = \frac{3}{7}$. Similarly,

$$6 \times \tfrac{4}{29} = 6 \times 4(\tfrac{1}{29}) = 24(\tfrac{1}{29}) = \tfrac{24}{29}.$$

In general terms $a \cdot \dfrac{b}{c} = \dfrac{ab}{c}$. However, since a fraction can be increased not only by multiplying the numerator but also by decreasing the denominator another method of multiplying fractions is possible. Two times one-quarter is the same as one times a half ; or $2 \cdot (\frac{1}{4}) = \dfrac{1}{4 \div 2} = \dfrac{1}{2}$, which is of course the same as $\frac{2}{4} = \frac{1}{2}$. Written in general terms

$$a \cdot \left(\dfrac{b}{c}\right) = \dfrac{b}{c \div a}.$$

This second rule for multiplication of fractions is used in the " cancelling method." For example, $9 \times \tfrac{5}{27}$ can be written as $\dfrac{9 \times 5}{9 \times 3} = \dfrac{5}{(9 \times 3) \div 9} = \dfrac{5}{3}$, which is the same result as cancelling the 9's would have given.

To multiply fractions by fractions we have to multiply all the numerators together and then all the denominators. For example

$$\dfrac{a}{b} \times \dfrac{c}{d} \times \dfrac{e}{f} = \dfrac{ace}{bdf}$$

or using numbers

$$\tfrac{1}{3} \times \tfrac{5}{4} \times \tfrac{2}{7} = \tfrac{10}{84} = \tfrac{5}{42}.$$

In this last example we might have done the cancelling before working out the multiplication. Any numerator can be cancelled with any suitable denominator as all the numerators are factors of the new numerator (in the result) and all the denominators are factors of the new denominator. But there is no need to stress this as it is one of the earliest things we learn in arithmetic.

We have now only division with fractions and powers of fractions to discuss. We will take the latter first because the raising to a power is only a kind of multiplication.

$$\left(\frac{a}{b}\right)^3 \text{ stands for } \frac{a}{b} \times \frac{a}{b} \times \frac{a}{b} = \frac{a \times a \times a}{b \times b \times b} = \frac{a^3}{b^3}.$$

The rule is simple enough : both the numerator and the denominator are raised to the same power as the whole fraction.

Of the simplest kind of division of fractions there is nothing new to say. If the dividing number is a factor of the numerator we just divide it into the numerator. For example,

$$\tfrac{6}{7} \div 3 = \tfrac{2}{7}.$$

The matter is rather more complicated if we are asked to divide one fraction by another, or a whole number by a fraction. We cannot immediately see the answer to $5 \div \tfrac{3}{4}$. The result will doubtless be bigger than 5 because $\tfrac{3}{4}$ is less than 1—but how much bigger ?

Like the ancient Egyptians and Greeks, we shall have to take refuge in unit fractions, but we shall use them in a different way. We will work it out like this ; supposing we were asked to divide 30 by 15 we could write it down as $30 \div (5 \times 3)$. Whether we divide first by 5 or by 3 we shall in any case get the answer 2 after dividing by both 3 and 5. Similarly with our fractions ; in dividing 5 by $\tfrac{3}{4}$ we think of it as $5 \div (3 \times \tfrac{1}{4})$ and we will first divide by the quarter. Since there are 4 quarters in 1 there will be 20 in 5, so that now dividing by the 3 we get the answer $\tfrac{20}{3}$ or $6\tfrac{2}{3}$. Here is another example worked out in the same way :—

$$7 \div \tfrac{5}{9} = 7 \div (5 \times \tfrac{1}{9}) = (7 \times 9) \div 5 = \tfrac{63}{5} = 12\tfrac{3}{5}.$$

We can state this method of division in general terms :—

$$\text{Dividend} \div \tfrac{\text{Numerator}}{\text{Denominator}} = \text{Dividend} \div \left(\text{Numerator} \times \tfrac{1}{\text{Denominator}} \right) \text{ (1)}$$

$$= \frac{\text{Dividend} \times \text{Denominator}}{\text{Numerator}} \quad \cdots \cdots \text{ (2)}$$

Notice that line (1) of this statement can be written

$$= \text{Dividend} \times \tfrac{\text{Denominator}}{\text{Numerator}}.$$

We began with a division by a fraction in line (1) and ended in line (2) with a multiplication by a different fraction. The second fraction however is merely the first one turned upside down. It is called " *the reciprocal* " of the first ; the numerator and denominator have changed places. The reciprocal of $\frac{3}{4}$ is $\frac{4}{3}$, of $\frac{5}{9}$ is $\frac{9}{5}$ and of $\frac{a}{b}$, $\frac{b}{a}$.

In this explanation of division by a fraction we assumed that the dividend was a whole number, but it makes no difference to the method we use if the dividend is not a whole number. In either case we turn division by a fraction into multiplication by the reciprocal of the fraction. In mathematical shorthand

$$n \div \frac{a}{b} = n \times \frac{b}{a}$$

whatever n is, whether algebraic or numerical, positive or negative, integral or fractional. Therefore it is also correct to write

$$\frac{n}{m} \div \frac{a}{b} = \frac{n}{m} \times \frac{b}{a} = \frac{nb}{ma}:$$

Now let us apply our new rule to some of our previous examples :—

$$5 \div \tfrac{3}{4} = 5 \times \tfrac{4}{3} = \tfrac{20}{3} = 6\tfrac{2}{3}$$
$$7 \div \tfrac{5}{9} = 7 \times \tfrac{9}{5} = \tfrac{63}{5} = 12\tfrac{3}{5}.$$

So our method works. Here are some more examples to illustrate it :—

$$\tfrac{5}{3} \div \tfrac{7}{8} = \tfrac{5}{3} \times \tfrac{8}{7} = \tfrac{40}{21} = 1\tfrac{19}{21}$$
$$\tfrac{17}{18} \div \tfrac{5}{9} = \tfrac{17}{18} \times \tfrac{9}{5} = \tfrac{17}{2} \times \tfrac{1}{5} = \tfrac{17}{10} = 1\tfrac{7}{10}.$$

Our method also enables us to divide any fraction by any whole number. We multiply the fraction by the reciprocal of the whole number, thus :—

$$\frac{a}{b} \div c = \frac{a}{b} \div \frac{c}{1} = \frac{a}{b} \times \frac{1}{c} = \frac{a}{bc}.$$

We can extend the application of our rule to the case in which both dividend and divisor are whole numbers :—

$$a \div b = \frac{a}{1} \div \frac{b}{1} = \frac{a}{1} \times \frac{1}{b} = \frac{a}{b}.$$

Thus we have turned a straightforward division sum into a multiplication sum. Here is an example with numbers :—

$$100 \div 25 = \tfrac{100}{1} \times \tfrac{1}{25} = \tfrac{100}{25} = 4.$$

This of course does not make the calculation any easier ; it is just another way of looking at it.

The rules we have arrived at for the working out of problems with fractions can be universally applied. We have only outlined the principles and do not suggest that we have covered the whole ground, but we have done enough to enable us to begin an examination of a subject already referred to, namely equations.

CHAPTER XIII

EQUATIONS

WE have already mentioned " the unknown x " and the important part it plays in equations. Methods of solving equations, that is to say of working them out, provide almost automatically the answer to some puzzles. The equals sign as used in equations does not indicate a statement of what *is* but gives a command to *make* equal. The command tells us to find a value for x which, as it is said, " satisfies the equation " or establishes the equality.

Without spending more time on theory we will set ourselves a puzzle. We will ask ourselves, " What is the number x, completely unknown to us, which satisfies the following conditions : if it is multiplied by 7 and 15 is then added, it gives us a number 10 times the original number ? "

Writing this mathematically it looks like this :—

$$7x + 15 = 10x.$$

We could begin to solve the puzzle or rather the equation by substituting numbers for x, beginning with 1 and continuing in a methodical way with 2, 3, 4. . . . When we reach the number 5 we find that this gives us the equality

$$35 + 15 = 50.$$

The problem is solved. The number x which we want is 5. But this method has little to recommend it. It certainly worked in this example but we shall rarely find our answer by it as easily as in this *chosen* example. The number x in any other problem might be a fraction, it might be negative, it might be very large indeed and there might be many other numbers which satisfied the conditions.

Before we examine sounder methods of solving equations there is a point to be noticed. In an equation the sign $=$ should be read not as " equals " but as " is to be equated to." We are warned that something active is required of us.

As we have decided not to solve our problems by trial and error, there remains another possibility—that of arriving at a solution by calculation and by some necessary deduction. If we could succeed in getting all the xs on one side of the equation and transferring all the other quantities to the other side we ought in the end to get an equation of this kind : $x = a$ or $nx = b$. Could we but get as far as this, our problem would be solved. For $x = a$ is exactly what we are looking for. If we ended up with $nx = b$, then we could compare this with the numerical example $2x = 6$; we know the answer to that, $x = \frac{6}{2} = 3$; therefore in the same way, if $nx = b$, then $x = \frac{b}{n}$.

It is all very well to be cocksure once we have reached this final stage of our calculation, but we should find in practice that the real difficulties arise on the way whilst we are attempting to separate our xs from other quantities. We need a reliable and universal method for getting to the final stage. In the historical development of algebra it took many hundreds of years before mathematicians achieved certainty in dealing with algebraic equations. Work on elementary equations was not completed until the sixteenth and seventeenth centuries.

There are some important points concerning equations which we ought to stress. In every equation there are two completely different kinds of quantities, the known constants and the unknown x (we are at present only concerned with equations in which there is one unknown). The whole understanding of what takes place in solving an equation depends on realizing the fundamentally different character of the two kinds of quantities. We will illustrate this important difference between the two quantities by some examples.

In the equation we have already examined

$$7x + 15 = 10x.$$

Of the three terms, 15 is the constant, $7x$ and $10x$ are multiples of the unknown. Alternatively we could say that the unknown x occurs with the two coefficients 7 and 10. Suppose we now look at a different equation,

$$7x + 9x - 2x + 5 = 33.$$

Here we can collect together all the xs by adding or subtracting the coefficients according as the signs tell us. The equation then becomes in a simpler form

$$14x + 5 = 33,$$

though this is of course not yet a solution. In the first example the xs are separated by the equals sign but the

constants are together. In the second example the constants
are separated and the xs are together. In neither case can we
immediately find the solution to the equations because the
equals sign appears to block the way. We are prevented from
writing them immediately in the form $nx = b$. Nor do our
difficulties end here. We might have equations composed
entirely of letters as indeed most mathematical equations are.
For example, $$nx + a = b - mx.$$

At this point our power of imagination fails us completely.
We must accept the fact that a, b, n and m are known from
the outset ; the only requirement is that x must be expressed
in terms of these constants, that is to say

$$x = \frac{b - a}{m + n}$$

which by the way is the solution of our equation. How in the
world do we get this solution ? How do we know that it is the
only one which will make the equals sign true by equating
both sides of the equation ?

FIG. 10.

We need a picture to help us. Let us imagine a pair of
scales. In the pans we can place knowns and unknowns just
as we wish: The scales are balanced when the pans are
equally loaded : the equation is balanced when one side equals
another. The equals sign of the equation corresponds to the
fulcrum of the scales. If we tamper with the contents of the
scale pans we have to balance things up again. In the same

way, if we alter the composition of the sides of an equation we have to balance things up again and make both sides equal. Still keeping to our analogy of the scales, let us think of the constants as ounce weights and the unknowns as apples all of equal though unknown weight. The solution of our equation is the same thing as finding the weight of one of the apples when apples and weights are mixed up together on both pans with the scales balanced. Let us draw a picture to represent the equation

$$2x + 15 = 3x + 3.$$

On one pan there are 2 apples and 15 ounce weights, and on the other 3 apples and 3 ounce weights. The problem is how to find the weight of one apple by exchanging or removing both apples and weights until we have only one apple remaining on one side and balancing weights on the other. At the same time we have to keep our eye on the pointer of the scales. It must not be allowed to move away from the " balanced " position except possibly for a moment whilst we are manipulating weights and apples. This keeping of the balance corresponds to fulfilling the command of the equals sign in the equation.

We will begin by trying to clear one scale pan of all its weights. If we take 3 ounce weights from both pans at the same time the balance will be maintained except for a momentary disturbance if we are heavy handed. The picture will now look like this :—

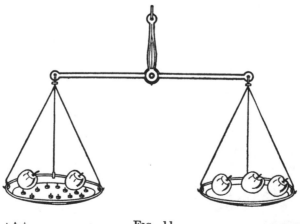

FIG. 11.

Expressed mathematically this picture has now become
$$2x + 12 = 3x.$$
We only need to remove the two apples from the left-hand scale pan to leave it occupied by weights only. To maintain the balance, however, we must at the same time remove two apples from the right-hand pan. The picture now looks like this :—

FIG. 12.

The apple evidently weighs 12 oz., or mathematically
$$x = 12.$$
Now let us check this result. Our original equation was
$$2x + 15 = 3x + 3.$$
If we replace x by the number 12 wherever there is an x we get
$$2x + 15 = 2 \times 12 + 15 = 24 + 15 = 39$$
and in the same way
$$3x + 3 = 3 \times 12 + 3 = 36 + 3 = 39.$$
In other words, the order to make both sides of the equation equal is carried out when x is equal to 12.

Just by looking at a picture we have arrived at an important rule. In fact the picture is hardly necessary ; the rule is almost self-evident. Two equal quantities will remain equal if we add to each of them, or subtract from each of them, other equal quantities.

We could develop our picture of the balance but it is hardly necessary to draw one each time. We will rely upon our imagination. If one apple is balanced by 12 ounce weights, then it should be clear that 7 apples, each of the same weight, would be balanced by 84 ounce weights. The balance would be held if we divided both sides by the same number. Extending this idea to equations and putting more trust in our imagination, if we raise one side of our equation to a certain power then the other must be raised to the same power in order to keep the balance or state of equality. We must be quite certain of what we mean by " the side " of an equation. The equals sign divides an equation into the left-hand and the right-hand sides. For example, in the equation

$$5x + 12 = 2x + 8$$

$(5x + 12)$ is the left-hand side and $(2x + 8)$ is the right-hand side.

The important thing to remember from all this is that whatever operation is carried out on one side of an equation, the same operation must be carried out on the other side if the state of equality is to be maintained. In practice this principle gives rise to the well-known " rule of thumb " for dealing with equations : move quantities from one side to another and isolate the unknown x, remembering to change the sign if you change the side.

Suppose we begin with the equation

$$5x + 12 = 2x + 8.$$

We will act as we did with our scale pans and try to assemble all the constants on one side, all the xs on the other. In order to do this we will first take away $2x + 8$ from both sides :—

Subtracting
$$
\begin{aligned}
5x + 12 &= 2x + 8 \\
2x + 8 &= 2x + 8 \\
\hline
3x + 4 &= 0
\end{aligned}
$$

Something alarming has happened. $3x + 4$ is equal to 0. Does this mean that we can go no further ? Does it mean that x is equal to 0 ? We must remember that we have not yet finished our job. We have not yet separated the knowns and

the unknowns. Let us see what happens when we take 4 from
both sides :—

Subtracting

$$3x + 4 = 0$$
$$4 = 4$$

$$3x = 0 - 4$$

Now we have $3x = -4$ and the 0 at least no longer worries
us because it has disappeared. We have almost solved the
equation. All we have to do is to isolate the x by dividing the
left-hand side by 3. Remembering our rule, however, we
shall have to divide the right-hand side by 3 at the same time
in order to keep our equation balanced. So

$$3x \div 3 = (-4) \div 3$$

or

$$x = (-4) \div 3 = -\tfrac{4}{3} = -1\tfrac{1}{3}.$$

The equation is solved. Now let us see if it is correct. To do
this we substitute $-1\tfrac{1}{3}$ or $-\tfrac{4}{3}$ for x in the original equation :—

The left-hand side is

$$5x + 12 = 5 \times (-\tfrac{4}{3}) + 12 = -\tfrac{20}{3} + 12$$
$$= -6\tfrac{2}{3} + 12$$
$$= 5\tfrac{1}{3}.$$

The right-hand side is

$$2x + 8 = 2 \times (-\tfrac{4}{3}) + 8 = (-\tfrac{8}{3}) + 8$$
$$= -2\tfrac{2}{3} + 8$$
$$= 5\tfrac{1}{3}.$$

So we can see that both sides are equal and our check has
verified our solution.

As we now understand the principles on which the rule of
thumb methods are based, we will use them in solving the
same equation, $5x + 12 = 2x + 8$. First we take $2x$ from
both sides. We remove the xs from the right-hand side completely by doing this :—

$$5x - 2x + 12 = 8$$

or

$$3x + 12 = 8.$$

Next the unwanted 12 must be disposed of. To do this we
subtract 12 from both sides and leave the left-hand side occupied by xs only.

$$3x = 8 - 12$$

or

$$3x = -4.$$

To find the value of one x we have to deal with the 3 on the
left-hand side. Since this 3 stands as a multiplying factor, it

can only be removed by division, so if we divide both sides by 3 we are left with $x = -\frac{4}{3} = -1\frac{1}{3}.$

We have completed the solution and have established the rule that if a term is moved from one side of an equation to the other, the operational sign attached to it must be replaced by the opposite sign, *i.e.*, addition must be changed to subtraction, subtraction to addition, multiplication to division, division to multiplication.

We have learnt how to deal with " linear " equations, that is with equations in which the unknown is raised only to the first power. Why the word linear is used in such a sense will be explained later in a geometrical way.

The reader can, if he wishes, work out all the examples we have mentioned so far, using the rules of thumb. We will show how these rules are applied in more difficult exercises, showing every step but making no comments. First :—

Solve
$$5(x - 2) - 2x = 2(x - 1)$$
$$5x - 10 - 2x = 2x - 2$$
$$3x - 10 = 2x - 2$$
$$3x - 2x = +10 - 2$$
$$x = +8$$

Secondly :—

Solve
$$a(b - c) - b(a + c) = ab - (bc - x)$$
$$ab - ac - ab - bc = ab - bc + x$$
$$-ac - bc = ab - bc + x$$
$$-ac - bc - ab + bc = x$$
$$-ac - ab = x$$

or
$$x = -ab - ac$$
$$= -a(b + c).$$

Notice in the second example that near the end we changed the *whole* equation round without changing the signs. This is permissible because if $x = 5$ it is just as true to say that $5 = x$ on account of the nature of the equals sign.

Supposing we had obtained an answer like this :—
$$-x = -10,$$

that is, with minus signs on both sides. We can multiply both sides by -1 and get the result
$$(-x) \times (-1) = (-10) \times (-1)$$

or
$$x = 10.$$

It is usual to multiply by -1 in this way whenever $-x$ arises at the end of a solution because it is generally $+x$ that we are looking for.

Sometimes equations at first glance look as though they could not be solved by our rules because they contain x in powers higher than the first, for example

$$(x + 1)(x - 1) = x^2 + x + 1.$$

This equation contains x^2 and because of this it is apparently a so-called " quadratic " equation. If we simplify it, however, it becomes a harmless linear equation which we know how to solve.

$$x^2 + x - x - 1 = x^2 + x + 1$$
$$x^2 - x^2 - x = 1 + 1$$
$$-x = 2$$
$$(-x) \times (-1) = 2 \times (-1)$$
$$x = -2.$$

The next equation looks rather difficult too. It is

$$\frac{1}{x - 1} - \frac{1}{x + 1} = 2.$$

Here x occurs only in the denominator. We will solve the equation step by step. The first thing to do is to put the two fractions over a common denominator.

$$\frac{1 \cdot (x + 1) - 1(x - 1)}{(x - 1)(x + 1)} = 2$$

$$\frac{x + 1 - x + 1}{x^2 - x + x - 1} = 2$$

$$\frac{2}{x^2 - 1} = 2 \quad \text{which is the same as}$$

$$2 \div (x^2 - 1) = 2$$

So
$$2 = 2 \times (x^2 - 1)$$
$$2 = 2x^2 - 2$$
$$2 + 2 = 2x^2$$
$$4 = 2x^2$$
$$2 = x^2$$

or
$$x^2 = 2.$$

This last is actually a quadratic equation the solution of which we are not yet able to complete because we do not yet under-

stand how to find " roots " of numbers. The final result
would be $x = \pm \sqrt{2} = \pm 1{\cdot}414 \ldots$

A further example is

$$\frac{5 + x}{2} + \frac{13 + 7x}{3} = 72$$

$$\frac{3 \cdot (5 + x) + 2 \cdot (13 + 7x)}{2.3} = 72$$

$$3 \cdot (5 + x) + 2 \cdot (13 + 7x) = 72 \times 2 \times 3$$
$$15 + 3x + 26 + 14x = 432$$
$$17x = 432 - 15 - 26$$
$$17x = 391$$
$$x = \tfrac{391}{17} = 23.$$

Here are some general observations : the solution of equations is one of the most important concerns of mathematics
and there are an untold number of tricks and dodges for
making the solution easier. Every mathematician must make
himself so familiar with the methods of solving equations that
he transfers terms, changes signs and isolates the unknown x
as though in a dream. Every text-book contains a large
number of well-chosen examples. The reader cannot be too
strongly advised to solve as many of these as possible, and
those of his own invention too. Solving equations is quite as
amusing as doing crossword puzzles or playing cards. He will
acquire that sixth sense which always astounds the layman
when he meets it in a mathematician. A mathematical sense,
or turn of mind, is almost a physical attribute, like being able
to swim, to ride a bicycle or to play the right stroke at tennis
without having to think about it. This sense can only be
acquired with practice except in the very highest sphere of
mathematics which hardly concerns us here. Our aim is
simply to become good general practitioners.

Before we extend our knowledge of types of equations let
us make use of what we already know in order to solve some
puzzles. Here is an example :—

A father is 48 and his son 21 years old. How many years
ago was the father's age exactly 10 times that of his son ?
And how old were they both then ? This is how we can solve
the puzzle. Suppose it was x years ago that the father's age
was 10 times that of his son. At that time the father's age

would have been $(48 - x)$ years and the son's age $(21 - x)$
years ; as the father's age was 10 times that of his son

$$(48 - x) = 10 (21 - x).$$

We have now expressed the problem as an equation. We
need worry no more about fathers, sons and their ages ; we
need only give our mind to the solution of the equation.

$$48 - x = 210 - 10x$$
$$10x - x = 210 - 48$$
$$9x = 162$$
$$x = 18$$

Eighteen years ago the father's age was $(48 - 18)$ or 30 years
and his son's age was $(21 - 18)$ or 3 years. As 30 equals
10×3, it is clear that we have the solution of the puzzle.

We could have arrived at it in another way. The father is
always $(48 - 21)$ or 27 years older than his son. Suppose
the son's age was x years when his father's age was 10 times
as great as his. The father's age must then have been $10x$
years. It must also have been $(27 + x)$ years, because he is
always 27 years older than his son. So $10x$ and $(27 + x)$
must be equal numbers :—

$$10x = 27 + x$$
$$9x = 27$$
$$x = 3.$$

The son was 3 years old and the father 30 ; and these were
their ages $(48 - 30)$ or 18 years ago.

We have solved the puzzle in two different ways and not
unnaturally have the same answers for both. The two
methods (and there may well be others) have been given to
show that even in this simple case the mathematician has a
choice of method. It is in this choice of method that he can
display his mathematical agility ; the possibility of choice
gives him the opportunity of selecting the simplest or perhaps
the most elegant method.

We will end this chapter with a puzzle which is famous and
ancient, the " Epitaph of Diophantus." Diophantus was a
mathematician who lived in Alexandria in the fourth century
A.D. He was a gifted writer on arithmetic and algebra,
perhaps the only outstanding one that the Greeks produced,
for almost all other Greek mathematicians were primarily
geometers. It was principally Diophantus who extended

knowledge of equations and who greatly influenced Arab and mediæval mathematicians. The epitaph inscribed upon his tomb runs (more poetically than we attempt here) :—

"Here lies Diophantus—a wonder to behold ; the stone tells his age by means of his art. He was granted a sixth of his life for his childhood ; it took him a further twelfth to grow his beard ; after one-seventh more he married and five years later he was presented with a son. The happy child, so well-beloved, reached half his father's age and then died. After four years of sorrow mitigated by the pleasures of mathematics, he himself at last reached his end."

In sober mathematics this epitaph becomes step by step an equation in algebra. We want to know the age at which Diophantus died ; this is the unknown, the " x " of the equation. Then $\frac{x}{6}$ years is the length of his childhood ; for $\frac{x}{12}$ years he grew his beard ; for a further period of $\frac{x}{7}$ years he was a young bachelor ; after 5 years of married life his son was born ; his son lived for $\frac{x}{2}$ years ; and the father survived 4 years more, dying at the ripe old age of x years.

We can now write down the equation

$$\frac{x}{6} + \frac{x}{12} + \frac{x}{7} + 5 + \frac{x}{2} + 4 = x.$$

The Lowest Common Multiple of the denominator is 84, so both sides of the equation are multiplied by 84.

$$14x + 7x + 12x + 420 + 42x + 336 = 84x$$
$$75x + 756 = 84x$$
$$756 = 9x$$
$$9x = 756$$
$$x = \tfrac{756}{9} = 84.$$

Diophantus therefore lived for 84 years. He was a boy for 14 years, by the age of 21 he had " grown his beard," he married at 33 and his son was born when he was 38. The son died at the age of 42 when his father was 80 years old and the father died 4 years later when he had reached the age of 84.

CHAPTER XIV

INDETERMINATE EQUATIONS

WE did not conjure up the ghost of Diophantus in the last chapter for nothing. This great mathematician is usually held to be the discoverer of a particular kind of equation, the indeterminate or Diophantine equation. How far this discovery can be attributed to Diophantus is as indeterminate as " his " equations. There is nothing in those of his writings which have survived to suggest to historians of mathematics that he was the originator of the equations to which his name has been given. But it is customary to use the name so we will keep it.

What are these puzzling Diophantine equations ? This time let us begin with a problem and discover both the nature and the solution of these extremely important equations as we go along. Here is the problem : find 2 numbers which are such that 8 times the first added to 3 times the second makes a total of 91.

Our present knowledge is obviously insufficient to enable us to solve this for we can see that *two* unknown quantities are involved here ; let us call them x and y. We therefore write

$$8x + 3y = 91.$$

What are we to do next ? According to our rules

$$8x = 91 - 3y$$

and

$$x = \frac{91 - 3y}{8}$$

or

$$3y = 91 - 8x$$

and

$$y = \frac{91 - 8x}{3}.$$

Clearly we can get no further for we are only expressing one unknown in terms of another unknown. A lucky shot would have enabled us to put x equal to 5 and y equal to 17. Then $8 \times 5 + 3 \times 17 = 40 + 51 = 91$. That would be a solution but we suspect that it is not the only one so we must seek a systematic method of finding all possible answers. Before doing so we must mention an important restriction, that is that the solutions to Diophantine equations must be whole numbers, otherwise we should find an endless number of solutions. We should only have to replace x by any number, say 7, and the equation would become

$$8 \times 7 + 3y = 91$$
$$56 + 3y = 91$$
$$3y = 91 - 56$$
$$3y = 35$$
$$y = \tfrac{35}{3} = 11\tfrac{2}{3}.$$

If we did this we should make the equation one in which there is only one unknown and we could therefore solve it immediately. If we put x or y equal to a number we are making that x or that y constant for the particular equation.

If we restrict the solutions to whole numbers for both unknowns occasions will arise when no solution at all can be found. But of this more later.

Before we can describe the general method of solution which the genius of the mathematician Euler evolved, we must become better acquainted with a very important mathematical dodge which is known as " substitution."

Let us take a concrete example. It can hardly be said that the equation

$$2\left(\frac{x-4}{3}\right) + 3\left(\frac{x-4}{3}\right) - 4\left(\frac{x-4}{3}\right) = 9 - 2\left(\frac{x-4}{3}\right)$$

looks very encouraging. Upon closer examination we notice however that the expression $\left(\dfrac{x-4}{3}\right)$ keeps recurring and, moreover, that x occurs only within this expression. We can now consider this expression $\left(\dfrac{x-4}{3}\right)$ to be a new generalised quantity, that is to say, a new unknown to which we must give a name. Let us call it n.

The equation can now be written

$$2n + 3n - 4n = 9 - 2n.$$

For the present the meaning of n is pushed into the background. We will worry more about its meaning when we know its numerical value. Our new equation with n as the unknown can be solved in the usual way :—

$$2n + 3n - 4n + 2n = 9$$
$$3n = 9$$
$$n = 3.$$

Now, we know that $n = \left(\dfrac{x-4}{3}\right)$ because we ourselves made it so. As n is no longer unknown but is a constant equal to 3, we can write down a new equation :—

$$n = \frac{x-4}{3}$$

from which $x = 3n + 4$. As $n = 3$, $x = 3 \times 3 + 4 = 9 + 4 = 13$. A check to ensure that this answer is in fact correct can be undertaken by the reader if he wishes.

We have seen here a practical example of substitution. We replaced the complicated expression $\left(\dfrac{x-4}{3}\right)$ by the simpler expression n wherever $\left(\dfrac{x-4}{3}\right)$ occurred. At our stage of learning this is the usual procedure, but in higher branches of mathematics, and particularly in integration, it sometimes happens that a simple expression is replaced by a more complicated one. In general, therefore, substitution means the replacement of one expression by another. There is usually little choice, and the motive for substitution is to simplify—in the end if not immediately.

We must return to the solution of Diophantine equations, using a special kind of substitution for it. But we shall soon find that other complications arise.

Euler's method of solving these equations demands that first we express one unknown in terms of the other. These are good practical grounds for expressing the unknown with the smaller coefficient in terms of the other. Our original equation was

$$8x + 3y = 91 \quad . \quad . \quad . \quad . \quad . \quad . \quad \text{(i)}$$

so that we want to express y in terms of x. Therefore we write

$$3y = 91 - 8x$$

$$y = \frac{91 - 8x}{3} \quad . \quad . \quad . \quad . \quad . \quad . \quad \text{(ii)}$$

This is the first step. Following Euler's method we handle this last expression as though it were an improper fraction. In this case we must divide by the 3 and separate the whole numbers from the fractions :—

$$y = \frac{91}{3} - \frac{8x}{3}$$

$$= 30 + \tfrac{1}{3} - 2x - \frac{2x}{3}$$

$$= 30 - 2x + \tfrac{1}{3} - \frac{2x}{3}$$

$$= 30 - 2x - \frac{2x - 1}{3}.$$

There is a small point here which should be mentioned. The fraction $-\left(\dfrac{2x-1}{3}\right)$ could equally well have been written as $+\left(\dfrac{1-2x}{3}\right)$. We shall find that the first form is the more convenient in use.

Now the calculation proper begins. We must remember that in Diophantine equations both y and x must be whole numbers. In our equation above, the left-hand side, y, must be a whole number. On the right-hand side 30 is a whole number, $2x$ also. The only doubtful expression is $\dfrac{2x - 1}{3}$.

Now the substitution is made : boldly we say that $\dfrac{2x - 1}{3}$ must be a whole number and we will give it a name, say n_1. So we make the substitution

$$n_1 = \frac{2x - 1}{3}$$

remembering that n_1 is a whole number. We now turn this round so that x is expressed in terms of n_1 :—

$$n_1 = \frac{2x - 1}{3}$$
$$3n_1 = 2x - 1$$
$$2x = 3n_1 + 1$$
$$x = \frac{3n_1 + 1}{2} \quad \cdot \quad \cdot \quad \cdot \quad \cdot \quad \cdot \quad \text{(iii)}$$

Once again we repeat the process of separating the parts of this improper fraction.

$$x = \frac{3n_1}{2} + \tfrac{1}{2} = n_1 + \frac{n_1}{2} + \tfrac{1}{2} = n_1 + \frac{n_1 + 1}{2}.$$

Again we remember that both x and n_1 have to be whole numbers. Therefore $\dfrac{n_1 + 1}{2}$ must also be a whole number. Again we substitute to ensure that $\dfrac{n_1 + 1}{2}$ will always be a whole number and this time we give the name n_2 to the new substitute. So that $n_2 = \dfrac{n_1 + 1}{2}$ in which n_2 is a whole number.

Expressing n_1 in terms of n_2,

$$2n_2 = n_1 + 1$$

or $$n_1 = 2n_2 - 1 \quad . \quad . \quad . \quad . \quad . \quad . \quad \text{(iv)}$$

There are no more possible fractions to worry about and therefore no more new symbols are required. Our remaining task is to express both x and y in terms of the last n ; that is of n_2.

Let us begin with x. Equation (iii) tells us that

$$x = \frac{3n_1 + 1}{2}$$

and in this equation we express n_1 in terms of n_2 by using equation (iv), like this :—

$$\left. \begin{array}{l} x = \dfrac{3n_1 + 1}{2} \\[2mm] n_1 = 2n_2 - 1 \end{array} \right\}$$

therefore

$$x = \frac{3(2n_2 - 1) + 1}{2}$$

$$= \frac{6n_2 - 3 + 1}{2}$$

$$= \frac{6n_2 - 2}{2}$$

$$= 3n_2 - 1.$$

So much for x ; now for y. The best means we have of finding y is to return to equation (ii), namely

$$y = \frac{91 - 8x}{3}$$

and to substitute the $(3n_2 - 1)$ for x. Then

$$y = \frac{91 - 8(3n_2 - 1)}{3}$$

$$= \frac{91 - 24n_2 + 8}{3}$$

$$= \frac{99 - 24n_2}{3}$$

$$= 33 - 8n_2.$$

We have now found the general solution of the Diophantine equation $8x + 3y = 91$. We will write it down in a more compact form. As we decided that n_2 must be a whole number and as we need no longer distinguish between n_1 and n_2, we can use n instead of n_2 for the sake of simplicity. So the general solution is finally

$$\left. \begin{array}{l} x = 3n - 1 \\ y = 33 - 8n \end{array} \right\}$$

in which n can be any whole number we care to choose.

Let us see if Euler's method has in fact given us the solution. We can substitute for n any whole number, positive or negative (and we can include 0 if we wish). The equation was

$$8x + 3y = 91$$

If we choose to make n equal to 1, $x = 3 - 1 = 2$ and $y = 33 - 8 = 25$.

Then $\quad 8x + 3y = 16 + 75 = 91.$

If n equals 5, $x = 15 - 1 = 14$ and $y = 33 - 40 = -7$.

Then $\quad 8x + 3y = 112 - 21 = 91.$

If n equals -1, $x = -3 - 1 = -4$ and $y = 33 + 8 = 41$.

Then $\quad 8x + 3y = -32 + 123 = 91.$

If n equals -0, $x = 0 - 1 = -1$ and $y = 33 - 0 = 33$.

Then $\quad 8x + 3y = -8 + 99 = 91.$

And so on for as many other values of n we choose.

This is a truly remarkable result. To think that a simple formula will produce an endless number of whole-number solutions for the two unknown quantities !

In practice, when these Diophantine equations have to be solved there are usually some further restrictions on the solutions. For instance, the solutions might be limited to numbers lying between 1 and 100 or between — 10 and + 10, or to positive or to negative numbers. If a restriction of this kind is imposed it is usually quite easy to decide by making a few trials, what values of n will give the desired results.

We have introduced Diophantine equations as a means to an end, as we shall find out later. We will therefore not delve further into their behaviour however interesting this may be. We must not leave this subject with the impression that any equation of the form $ax + by = c$, in which a, b and c are whole numbers, is necessarily a true Diophantine equation, that is, one in which the two unknowns have integral (whole number) solutions. A further condition has to be satisfied before we can be sure of a whole number solution.

First of all we must see that the equation is expressed in the simplest possible way. If the coefficients have any common factors the equation must be divided by these factors. For example, the equation

$$9x + 12y = 51$$

can be simplified to

$$3x + 4y = 17$$

by dividing by the common factor 3. Similarly, the equation

$$32x + 24y = 124$$

can be simplified to become

$$8x + 6y = 31.$$

In this last equation we meet a difficulty. Whatever whole numbers we substitute for x and y both $8x$ and $6y$ are bound to be even numbers and the sum or difference of two even numbers can never be equal to an odd number. Yet in this equation the sum of the two even numbers is to be 31 : it is clear that no whole number solution is possible. It is not a true Diophantine.

Here is another deceptive-looking equation which can have no integral solution :—

$$3x + 15y = 19.$$

Whatever whole numbers we substitute for x and y, the left-hand side is bound to add up to a number divisible by 3. But as 19 is not divisible by 3 there is no possible solution in whole numbers.

We will try to express these two cases in a more general way and give a proof of the rule that covers them. First of all let us look at an example :—

$$9x + 12y = 51.$$

Dividing by 3, $3x + 4y = 17$.

The coefficients 3 and 4 have no common factor so this equation should be a true Diophantine with a whole-number solution. Let us apply Euler's method to it :—

$$3x + 4y = 17$$
$$3x = 17 - 4y$$
$$x = \frac{17 - 4y}{3} = 5 + \tfrac{2}{3} - y - \frac{y}{3} = 5 - y - \frac{y - 2}{3}$$
$$\frac{y - 2}{3} = n$$
$$y - 2 = 3n$$
$$y = 3n + 2 \; ; \; x = 5 - (3n + 2) - n = 3 - 4n.$$

In this example we did not need more than one n because, after putting $\dfrac{y - 2}{3}$ equal to n, no further fractions occurred. If $n = 3$, then $x = -9$ and $y = 11$; the equation is satisfied because

$$3 \times (-9) + 4 \times 11 = -27 + 44 = 17.$$

This example was solved without difficulty. Let us go back to the general equation and to our proof. We begin with $ax + by = c$ in which a, b and c are whole numbers which have no common factor. It might happen however that a and b had common factors as in the case of the equation

$$8x + 6y = 31,$$

in which both 8 and 6 are multiples of 2. Supposing, in our general equation, that a and b have a common factor which we will call m. Then the numbers $\dfrac{a}{m}$ and $\dfrac{b}{m}$ would also be whole numbers. If x and y are in whole numbers which satisfy the equation, that is to say, if they are solutions of the original equation, then $ax + by = c.$

As we know that $\frac{a}{m}x$ and $\frac{b}{m}y$ are whole numbers, their sum must also be a whole number. But their sum is equal to $\frac{c}{m}$ and this cannot be a whole number because we said that the number c was not divisible by m. So we meet a contradiction which can be overcome only if we agree that there can be no whole-number solutions of the equation $ax + by = c$ in this case. So we have proved that the equation which can be written in the form

$$m \cdot rx + m \cdot sy = c$$

in which m, r, s and c are whole numbers and m is not a factor of c, can have no solution in whole numbers.

Let us state this conclusion in another way : in a Diophantine equation the coefficient of the two unknowns must be "prime" to each other, that is they must have no common factor. It is of course possible for each of the two coefficients to have factors in common with the constant. For example, in the equation

$$3x + 4y = 12$$

3 and 12 have the common factor 3 ; 4 and 12 have the common factor 4. This equation has a solution, namely

$$x = 4 - 4n$$
$$y = 3n$$

If we put $n = 5$ (or any other number) to check this result, we see that $x = 4 - 20 = -16$ and $y = 15$. Substituting in the left-hand side of the equation

$$3 \times (-16) + 4 \times 15 = -48 + 60 = 12$$

and the solution is confirmed.

CHAPTER XV

NEGATIVE AND FRACTIONAL POWERS

IT is very tempting to continue our examination of equations because we should soon learn to look at them in a new way, that is, as " functions " as they are called. It will pay us however to exercise a little patience and extend our knowledge of the world of numbers for we shall find that this will enable us to move about in the world of functions much more easily.

The first step will be to refresh our memory of the manner of raising numbers to a power. It is a study of division which concerns us and which will yield us a further store of knowledge. We will use for the moment numbers expressed as powers of the same base. When we divide one such number by another we merely subtract the index of the divisor from the index of the dividend ; for example, $10^5 \div 10^3 = 10^{5-3} = 10^2$ or, in full, $100,000 \div 1000 = 100$. We said nothing at the time when we first discussed this subject, about the possibility of the index of the divisor being greater than that of the dividend. We dealt only with the case, expressed in a general way, of the division $a^m \div a^n$ where $m > n$. What is more, we always arranged that m and n would be positive numbers so that there was no chance of a negative index appearing. There is no reason why a negative index should not appear but we ought to see what such a thing would mean and how it would fit into the system of numbers which we have so far built up.

To begin with let us keep m and n as positive numbers but make the stipulation that n is to be greater than m $(n > m)$, with n attached to the divisor. If we carry out the division by the rule we already know, then $a^m \div a^n = a^{m-n}$; but something new arises here because if $n > m$ then the index $(m - n)$ must be negative. Here is a numerical example :—

if $\qquad\qquad a = 10, m = 5$ and $n = 7.$

then $\qquad\qquad 10^5 \div 10^7 = 10^{5-7}$

$$= 10^{-2}.$$

We can work more freely with numbers than with letters, so we will work out this result in detail :—

$$10^5 \div 10^7 = \frac{10 \times 10 \times 10 \times 10 \times 10}{10 \times 10 \times 10 \times 10 \times 10 \times 10 \times 10}$$

(cancelling) $= \dfrac{1}{10 \times 10}$

$= \dfrac{1}{10^2}.$

This result $\dfrac{1}{10^2}$ must be identical with 10^{-2}.

We may already have seen what is happening but we will take one more example to make sure :—

$$a^2 \div a^5 = \frac{a \times a}{a \times a \times a \times a \times a}$$

$$= \frac{1}{a \times a \times a}$$

$$= \frac{1}{a^3}.$$

This again should be identical with $a^{2-5} = a^{-3}$.

The rule is straightforward : a base raised to a negative power is equal to the reciprocal of the base raised to an equal positive power. The formula is $a^{-r} = \dfrac{1}{a^r}$ so long as a is not equal to 0.

The limitation that a must not be 0 is reasonable because if we put $a = 0$ in the formula, then $a^{-n} = \dfrac{1}{0^n} = \dfrac{1}{0}.$ We already know from Chapter XII that $\dfrac{1}{0}$ has no expressible value although we may rather loosely say that it is " infinite."

No more need be said about negative indices. We have seen how our rule incorporated them in our number system. We can manipulate expressions like $b^{(-4+3-2)}$ which equals $b^{(-3)}$ or $\dfrac{1}{b^3}$, just as easily as $c^{(2+4-3)}$ which we know is c^3.

We know that $a^0 = 1$ (see Chapter II), so we can write

$$\frac{1}{a^n} = \frac{a^0}{a^n} = a^{0-n} = a^{-n},$$

a result we already know. But we can use the same method
to find something more.

$$\frac{1}{a^{-n}} = \frac{a^0}{a^{-n}} = a^{0-(-n)} = a^n.$$

It is now possible to state a rule about reciprocals. The
reciprocal of a number raised to a positive power equals the
number raised to the same negative power, and vice versa.

Having brought negative indices into the system, let us try
to bring in fractional ones like $10^{\frac{2}{3}}$ or $a^{\frac{1}{2}}$ or $3^{\frac{2}{3}}$. Clearly it is
difficult to imagine what is meant by the command " $3^{\frac{2}{3}}$,"
that is to say " raise 3 to the power of $\frac{2}{3}$ " ; a^3 stands for
$a \times a \times a$ but how can we multiply a by itself " two-thirds
of a time " as $a^{\frac{2}{3}}$ would demand ? $3^{\frac{2}{3}}$ could be written $(3^2)^{\frac{1}{3}}$,
but does this help ? 3^2 is certainly 9, but what is $9^{\frac{1}{3}}$?

We shall have to give away the secret and indicate the new
idea which is involved in the understanding of fractional
powers. It is connected with what are called the " *roots* " of
a number. The number $9^{\frac{1}{3}}$ is the number, at present unknown,
which when multiplied by itself 3 times will give the number 9.
We cannot work this out yet as we are not yet equipped to
do so, but we shall return to this later.

Treating the subject of roots in a more general way, we
might consider the number $c^{\frac{a}{b}}$ which can be written $(c^a)^{\frac{1}{b}}$.
Suppose we make this number equal to d. Then

$$d = (c^a)^{\frac{1}{b}}.$$

If we raise both sides of this equation to the power b,

$$d^b = (c^a)^{\frac{b}{b}}$$
$$d^b = c^a.$$

If we put this into words we might say that d or $c^{\frac{a}{b}}$ is the
number which when raised to the power b is equal to c^a.

The reader is probably already familiar with the sign $\sqrt{}$
which is used in arithmetic to show that a root of a number
is required. This sign has been used for many centuries and
it appears to have been derived from the written letter " r,"
standing for " radix " (Latin, " root "). The letter " r "
placed in front of a number indicated that a root of the
number had to be extracted ; the extended line following it
was added to show clearly what numbers were to be included

in the operation. The relationship between the index form and the radix form of a number should now be clear ; $a^{\frac{1}{2}}$ is the same thing as $\sqrt[2]{a}$; they are two ways of writing " the square root of a." Similarly $3^{\frac{2}{3}} = (3^2)^{\frac{1}{3}} = \sqrt[3]{9}$, the " cube root " of 9, and so on.

Whenever calculations arise in which roots of numbers are involved it almost always pays to use the index form rather than the radix form for dealing with them. The index form is much simpler to manipulate. Here is an example :—

$$\sqrt[3]{a^2} \cdot \sqrt{a} = a^{\frac{2}{3}} \cdot a^{\frac{1}{2}} = a^{\frac{4}{6} + \frac{3}{6}} = a^{\frac{7}{6}},$$

which could be expressed in the radix form as $\sqrt[6]{a^7}$ if we like.

Negative fractional powers may occur too. This frequently happens when one number raised to a fractional power is divided by another as, for example, in the division $\sqrt[3]{a} \div \sqrt[2]{a}$ or $a^{\frac{1}{3}} \div a^{\frac{1}{2}}$. Here

$$a^{\frac{1}{3}} \div a^{\frac{1}{2}} = a^{\frac{1}{3} - \frac{1}{2}} = a^{\frac{2}{6} - \frac{3}{6}} = a^{-\frac{1}{6}}.$$

The result $a^{-\frac{1}{6}}$ could also be written either as $\dfrac{1}{a^{\frac{1}{6}}}$ or as $\dfrac{1}{\sqrt[6]{a}}$.
All these forms are equally acceptable though $a^{-\frac{1}{6}}$ is possibly the simplest and most convenient once we fully understand what the index means.

One further comment on the radix sign should be made before we leave it for higher realms of mathematics. When we have used it we have always indicated precisely which root we required by adding the corresponding number, e.g. 3, as the number to indicate the cube root in $\sqrt[3]{9}$. If the radix is used without an indicating number at all then it is the square root that is required ; so $a^{\frac{1}{2}}$ could be written $\sqrt{a}$ but $a^{\frac{1}{4}}$ would have to be written as $\sqrt[4]{a}$. If this rule is followed no mistake is likely to arise. The form $\sqrt[1]{a}$ is never used because $\sqrt[1]{a} = a^{\frac{1}{1}} = a^1 = a$.

Finally, let it be understood that the numerical methods of working out square roots and cube roots without using logarithms will not be described in this book. They can be found, if they are required, in most text-books of arithmetic. We are more concerned here to extend our idea of " number " than to describe in detail calculations which, however interesting in themselves, will not immediately further our progress.

CHAPTER XVI

IRRATIONAL NUMBERS

UNDER what conditions, considered from a purely general point of view, are we able to calculate the root of a number ? If we want to find $\sqrt[4]{a}$, for example, this is simple enough if we happen to know that a is equal to the number p^4. In that case

$$\sqrt[4]{a} = \sqrt[4]{p^4} = p.$$

We say that p is "*the fourth root*" of a.

Now let us see whether it is always as easy as this. Can we assume that we shall always find a number p raised to the fourth power, which equals a ? We will look at the easiest case and stipulate that a must be a whole number, any whole number we care to choose. Obviously we shall be very lucky if we can find the number p because amongst the first hundred positive whole numbers only 1, 16 and 81 are fourth powers. That is, if we restrict our choice to the numbers below 100 there are only these three which give whole-number fourth roots. The number 25, for instance, lies between 2^4 and 3^4 and the number 90 between 3^4 and 4^4, and so on.

There is an important use of familiar symbols which we will describe here because they are convenient for use in our present work. They are frequently employed in higher mathematics. We use the signs of inequality, " greater than " and " less than," the symbols $>$ and $<$, to write in a concise way that one number lies between two others. To say, for example, that the number b lies between 30 and 40, we could write

$$30 < b < 40$$

which is read as " 30 is less than b which is less than 40." If it is possible for the number b actually to be equal to 30 and to 40 then we write $\quad 30 \leqslant b \leqslant 40$

in the shorthand way of expressing this property of the number b. If we translate this symbolic statement as it stands, it reads " 30 is less than or equal to b which is less than or equal to 40." The numbers 30 and 40 are called the lower and upper " bounds " of b. If we meet the expression

$$a < b < c$$

then we know that the three numbers a, b and c are in increasing order, that a is the least and that c is the greatest.

We can now write the phrase " 25 lies between 2^4 and 3^4 " in shorthand :—
$$2^4 < 25 < 3^4.$$

It is obvious that the fourth root of 25 is not one of the type $\sqrt[4]{p^4}$ in which p is a positive whole number. However, the answer may be a fraction lying between 2 and 3. We know that there is an infinity of fractions lying between any two whole numbers. We can take fractions with denominators of 200, 2000 or 2 million if we like, so that it does look as though we stand a reasonable chance of finding the number p among these fractions. A rough guess at the value of p could be $2 + \frac{1}{4}$, which is the same as $\frac{9}{4}$. If $\frac{9}{4}$ is raised to the fourth power, $\left(\frac{9}{4}\right)^4 = \frac{9^4}{4^4} = \frac{6561}{256} = 25 \cdot 63. \ldots$ So our guess is not very wide of the mark. If we try some number which is a little less than $2\frac{1}{4}$ we might hit upon the correct value of p, though we might have to spend a long time at it.

Before we accept this argument we ought to analyse our problem in general terms. We know that there are two possible cases, one in which the nth root (n is any number) of a is equal to p^n where both a and p are positive whole numbers ; the other in which the value of a lies between two whole numbers, that is, we assume it will be a fraction. Let us call this fraction $\frac{r}{s}$ on the understanding r and s are " prime to each other "—this means that the fraction is in its lowest terms.

Stating this as an equation,
$$\sqrt[n]{a} = \frac{r}{s}.$$

Raising both sides to the power of n,
$$a = \left(\frac{r}{s}\right)^n.$$

In this last statement lies a contradiction. On the left-hand side we have " a," a whole number by definition. On the right-hand side we have the nth power of $\frac{r}{s}$. But $\frac{r}{s}$ is by definition a fraction in its lowest terms. Therefore the nth power of $\frac{r^n}{s^n}$ is also a fraction in its lowest terms. For, by

raising to a power we mean " multiplying by itself " ; we have
introduced no new factors and therefore no further simplification is possible.

This little proof has shown that the roots of whole numbers
are only occasionally other whole numbers. Furthermore,
when they are not whole numbers, they *cannot* be fractions.
So our hit-or-miss method of finding $\sqrt[4]{25}$ would never have
given us an exact answer however long we tried.

This is rather surprising. After all, there is an infinity of
fractions lying between any two adjacent whole numbers and
yet it is impossible to find one in this infinity to meet the quite
straightforward condition that it should equal $\sqrt[4]{25}$ or the
even simpler requirement $\sqrt{2}$.

To provide ourselves with an answer we extend our concept
of number and bring in a new idea. We say that there is another
class of numbers which pushes its way in between the fractions
(which we had previously defined as filling all the space between
one whole number and the next!) in an endless series.

When the Greek mathematician Pythagoras met this kind
of number he said they were " alogos," inexpressible, unreasonable ; we call them " irrational numbers." How are
these " inexpressible " numbers to be expressed ? Whole
numbers and fractions are banned ; what then ?

By logarithmic calculation we can show that $\sqrt[4]{25}$ is
2·23606 . . ., the dots indicating that the calculation is
incomplete. So this is an example of an " inexpressible "
number, the new class of number. As another example we
could take π, the number connected with circles. A rather
surprising equation was discovered by Leibniz :—

$$\frac{\pi}{4} = 1 - \tfrac{1}{3} + \tfrac{1}{5} - \tfrac{1}{7} + \tfrac{1}{9} - \tfrac{1}{11} \ldots \text{ and so on.}$$

This series of fractions continues without end. Though we
can calculate $\frac{\pi}{4}$ from the series to as many decimal places as
we like, we can never find the exact numerical value of $\frac{\pi}{4}$.

This suggests two ways of expressing irrational numbers.
We can use either decimal fractions or " infinite series," both
of which are essentially the same. For example, the Leibniz
series can be used to express the value of π as

$$3\cdot141592653589793 \ldots$$

Another very important irrational number, e, the basis of the natural logarithms can be written as the series

$$e = 1 + \frac{1}{1!} + \frac{1}{2!} + \frac{1}{3!} + \cdots$$

or alternatively as a decimal,

$$e = 2 \cdot 71828182845904523536 \ldots$$

The critical reader will have noticed that, contrary to our expressed intentions, we have taken something for granted in the previous paragraphs. We have used decimal fractions without explaining what they are. Agreed ; but we shall deal with them in the next chapter. However there is a more serious criticism to make. We said that irrational numbers cannot be expressed as exact fractions but that they could be represented as a never-ending decimal fraction. But it may be already known to the reader that some unending decimal fractions do in fact stand for exact common fractions ; $0 \cdot 33333 \ldots$ or $0 \cdot 3$ (3 recurring) is another way of writing the fraction $\frac{1}{3}$.

The distinction between rational (that is whole numbers and common fractions) and irrational numbers is not usually very well understood even by those who are familiar with them and use them in calculations. It is however an important distinction which ought to be grasped by all and that is why we have spent some little time over it here. A rational number can always be expressed exactly and completely by a simple fraction, but an irrational number can only be expressed numerically as a decimal fraction or as a series, neither of which can ever be completed. In other words, we can only get an approximate value for an irrational number though if we care to take the trouble we can usually make the approximation as close as we like by working to many places of decimals.

The reader will remember the recurring decimal fraction $\cdot 3333 \ldots$ which apparently never ends. Is this an irrational number ? No, because we know that this, like all other recurring decimals, can be represented exactly and completely by a simple fraction, namely $\frac{1}{3}$. We can point to the exact place on the " line of numbers " where this fraction lies. But $\sqrt[4]{25}$ or $2 \cdot 236 \ldots$ we could not locate on the line of numbers even if we used a microscope. Of course we can indicate rational numbers between which it must lie and we can make these numbers as close together as we like but because an irrational number is not expressible completely and exactly there will always be some uncertainty about its exact position.

Anyone who wishes to pursue the investigation of irrational numbers can find a fuller (but still elementary) explanation in " What is Mathematics ? " by Courant and Robbins, O.U.P., in which the work of Dedekind and Cantor is considered.

CHAPTER XVII

GENERALISED DECIMAL FRACTIONS

IT is certainly time that we looked more closely at decimal fractions. Although we shall refer to such fractions in other number scales we shall find it convenient to use the terms " decimal " and " decimal point " in places where the " deci " (ten) meaning of the terms is not strictly applicable. If we were working in the scale of two perhaps we ought to speak of the " binal " point. If we work in a perfectly general scale it is difficult to find a term more appropriate than " decimal " point.

In any case it will be apparent that if decimal fractions in the scale of ten are $\frac{1}{10}$, $\frac{1}{100}$, $\frac{1}{1000}$ . . . according to their place after the decimal point, then such fractions in the scale of 6 for instance will be $\frac{1}{6}$, $\frac{1}{36}$, $\frac{1}{216}$ . . .; or in the scale of 2, $\frac{1}{2}$, $\frac{1}{4}$, $\frac{1}{8}$, $\frac{1}{16}$. . . . In general terms therefore we could write a number in a general decimal system in this way :—

$$pb^2 + qb^1 + rb^0 + \frac{s}{b^1} + \frac{t}{b^2} + \frac{u}{b^3}$$

or alternatively

$$pb^2 + qb^1 + rb^0 + sb^{-1} + tb^{-2} + ub^{-3}.$$

Here b is the base of the scale, as 10 is of the decimal system proper. And p, q and r are the whole number coefficients, s, t and u are the coefficients of the fractions. The decimal point would come between rb^0 and sb^{-1}.

In our normal system (in the scale of 10) the number 341·732 stands for

$$3 \times 10^2 + 4 \times 10^1 + 1 \times 10^0 + 7 \times 10^{-1} + 3 \times 10^{-2} + 2 \times 10^{-3}.$$

This example should make clear the structure of numbers with decimal fractions.

As we are going to discuss in detail and in a general way further series like those of the previous paragraph, it will pay us to use symbols which will reduce these cumbersome expressions to a simpler form. Mathematicians always invent a shorthand whenever it looks as though it might be profitable. For writing series in which the terms change systematically and in order, from the first term to the last, we make use of the Greek capital letter " sigma," Σ, which we interpret as " the sum of all terms like. ...." For example, the series

$$x^1 + x^2 + x^3 + x^4 + \ldots\ldots$$

could be written Σx^r so long as we also understand that the index r is restricted to whole numbers. Our new symbol is not yet complete however. We might be interested only in those terms of the series which begin with x^3 and end with x^8. To show that these terms are the limits, we write the short-hand form as

$$\sum_{r=3}^{8} x^r$$

and read it as " the sum of all terms like x^r, in which r takes in turn the values of all the whole numbers from 3 to 8 inclusive." Two additional examples should make clear how the symbol is used and how much trouble it saves.

If the series $\dfrac{a}{1!} + \dfrac{a^2}{2!} + \dfrac{a^3}{3!} + \dfrac{a^4}{4!} \cdots \dfrac{a^{50}}{50!}$ is to be written using the Σ shorthand, the first step is to recognise the structure of the terms and how they vary systematically from one to the next. We then write down the " general term " as it is called ; that is, a completely algebraic version of a typical number of the series. In our example this would be written as

$$\frac{a^r}{r!}$$

using the letter r to stand for any number. The series could then be written as

$$\sum_{1}^{50} \frac{a^r}{r!}.$$

If the more general series

$$\frac{a_1 y^2}{1} + \frac{a_2 y^3}{2} + \frac{a_3 y^4}{3} \ldots \text{ for 16 terms}$$

is to be written using Σ, we look for the general term as we did in the previous example. In this case it will be $\dfrac{a_r y^{r+1}}{r}$ because we notice that the index of y is always one greater than the corresponding denominator. So the series would be written as

$$\sum_{1}^{16} \frac{a_r y^{r+1}}{r} \quad \ldots\ldots \quad \text{Example (i)}$$

In some types of series it is not quite so easy to express the general term. We have a new and rather important case in the Leibniz series for π which we have already quoted :—

$$\frac{\pi}{4} = \tfrac{1}{1} - \tfrac{1}{3} + \tfrac{1}{5} - \tfrac{1}{7} \cdots$$

Here there are two difficulties. In the first place the numbers jump by 2's and not by 1's and in the second place the signs are no longer all plus, they alternate between plus and minus. The denominator is simple to deal with. The series $2, 4, 6, 8 \ldots$ can be thought of as $2 \times 1, 2 \times 2, 2 \times 3, 2 \times 4 \ldots$ with the general term $2r$. If we take 1 away from each term of this series we obtain the series of denominators $1, 3, 5, 7 \ldots$ which can therefore be thought of as $(2 \times 1 - 1), (2 \times 2 - 1),$ $(2 \times 3 - 1) \ldots$ with the general term $(2r - 1)$. Disregarding the sign, the general term of Leibniz's series can therefore be written as $\dfrac{1}{2r - 1}$.

The alternating sign is dealt with by a mathematical trick. We use the results

$$(-1)^1 = -1$$
$$(-1)^2 = (-1) \times (-1) = +1$$
$$(-1)^3 = (-1) \times (-1) \times (-1) = -1$$
$$(-1)^4 = (-1) \times (-1) \times (-1) \times (-1) = +1$$

and so on ; we notice that the value of $(-1)^n$ is $+1$ when n is even and -1 when n is odd. If therefore we multiply any quantity by $(-1)^n$ all that happens to it is that its sign is changed if n is odd ; if n is even we have merely multiplied the quantity by 1 so that it is in fact unchanged. Here is a device which will meet our second difficulty. Multiply the general denominator $\dfrac{1}{2r - 1}$ by $(-1)^r$. We then get $\dfrac{(-1)^r}{2r - 1}$ for the general term. Let us try it. If we put $r = 4$ in this formula we ought to get the result $-\tfrac{1}{7}$; in fact we get $\dfrac{(-1)^4}{8 - 1} = +\tfrac{1}{7}$; the sign is wrong though the denominator is correct. This need not disturb us ; let us multiply our general term by another (-1) so that it becomes

$$\frac{(-1)^{r+1}}{2r - 1}.$$

Then,   when $r = 1$, we have $\dfrac{(-1)^2}{2-1} = \dfrac{+1}{1}$

„ $r = 2$ „   „ $\dfrac{(-1)^3}{4-1} = \dfrac{-1}{3}$

„ $r = 3$ „   „ $\dfrac{(-1)^4}{6-1} = \dfrac{+1}{5}$

„ $r = 4$ „   „ $\dfrac{(-1)^5}{8-1} = \dfrac{-1}{7}$

So we see that our formula is correct and we can write, for example, the first 50 terms of the Leibniz series as

$$\sum_{r=1}^{50} \frac{(-1)^{r+1}}{2r-1}.$$

It must be emphasised that this Σ is used only for series that are " *discrete* " as well as systematic. By " discrete " we mean that there are jumps from one term of the series to the next. We say that the series " *runs* " from the first term to the last but it would be more accurate to say that it " *hops.*" Because the series is systematic the hop from one term to the next is always of the same kind. In example (i) which we gave, notice that in order to get from any one term to the next we have to take three actions :—

(1) We increase the suffix of a by 1.
(2) We raise the power of y by 1.
(3) We increase the denominator by 1. And we have to do this at every stage of the series.

This idea of a series progressing by hops or jumps is a very important one in mathematics. We shall see later in the integral calculus that we shall modify the nature of the jump from term to term and at the same time replace the summation sign Σ by the new sign $\int$.

After this digression it is time to return to decimal fractions. How can we use our new notation to represent in a general way a number with four figures from the decimal point for example ? It could be written either as

$$\sum_{r=1}^{4} a_r \times \frac{1}{b^r}$$

or as

$$\sum_{r=1}^{4} a_r \times b^{-r}.$$

Writing these formulæ in full to make the notation clear, they stand for

$$a_1 \times \frac{1}{b^1} + a_2 \times \frac{1}{b^2} + a_3 \times \frac{1}{b^3} + a_4 \times \frac{1}{b^4}$$

or $a_1 \times b^{-1} + a_2 \times b^{-2} + a_3 \times b^{-3} + a_4 \times b^{-4}.$

If we are working in the scale of 10 there would be a_1 tenths, a_2 hundredths, a_3 thousandths and a_4 ten-thousandths.

We shall sometimes need to consider a decimal fraction of n terms where n is an unknown but finite number. Such a decimal would be written

$$\sum_{r=1}^{n} a_r \times b^{-r} = a_1 b^{-1} + a_2 b^{-2} + \ldots a_n b^{-n}.$$

We already know that some decimal fractions never end. How can we express these in general terms ? It is simple enough. We make the upper limit " infinity " (wh.ch is to say we have no upper limit at all). So that

$$\sum_{r=1}^{\infty} a_r b^{-r} = a_1 b^{-1} + a_2 b^{-2} + \ldots a_r b^{-r}. \ldots$$

The dots indicate that the series continues without end.

If we wish to represent in the most general way a number with integers and decimal places we shall have to change the limits again and write

$$\sum_{r=m}^{-\infty} a_r b^r$$

to stand for

$$a_m b + a_{m-1} b^{m-1} + \ldots a_1 b^1 + a_0 b^0 + a_{-1} b^{-1} + a_{-2} b^{-2} \ldots$$

The decimal point of this number (of course not written in the series) would come between the terms $a_0 b^0$ and $a_{-1} b^{-1}$; so this number has $(m + 1)$ figures before the decimal point and an infinity of decimal places after it.

Although we use this way of expressing numbers greater than 1 with decimal attachments, we shall for the most part be interested in numbers which are less than 1. In this last phrase there is an ambiguity which we should clear up at once. There are two possible ways of interpreting the phrase " less than 1," or, more exactly, " a is less than b." The fraction $\frac{1}{2}$ is certainly less than 1. But is it less than $- 1$? The answer is both yes and no. If in our mind's eye we look along " the line of numbers " the number 0 is certainly less than 1, the number -1 is certainly less than 0—so that in this sense $\frac{1}{2}$ *is not less* than -1. If however we disregard the sign, then $\frac{1}{2}$ *is* less than -1. But this is an unsatisfactory state of affairs to a mathematician so he introduces the idea of " *the absolute value* " of a number. For the present it is sufficient to say that the absolute value of a number is its face value independent of its sign. To show that we are speaking of a number in this way we write it between two vertical lines, thus, $\mid 5 \mid$. Of course the value of $\mid 5 \mid$ is 5 and so is the value of $\mid -5 \mid$. It is in the absolute sense that we shall be speaking of " fractions less than 1."

After all this preparation we can at last resume our attack on decimal fractions. First of all let us calculate the value of the fraction $\frac{3}{40}$. By division we find that this is equal to 0·075 exactly. It gives a finite number of decimal places. Similarly the fraction $\frac{4}{125}$ can be written 0·032 exactly, again with a finite number of decimal places. We all know that $\frac{1}{2} = 0\cdot5$ and $\frac{1}{5} = 0\cdot2$. We are tempted to make a general rule. The fraction which in its lowest terms can be written in a general way as $\frac{p}{q}$ can be expressed as a finite decimal fraction in the scale of 10 if the denominator q contains only the factors 2 and 5 (raised of course to any power). Notice that the denominators of the fractions we used as examples were all of this kind.

$$40 = 2 \times 2 \times 2 \times 5 = 2^3 \times 5^1$$
$$125 = 5 \times 5 \times 5 \quad\;\; = 2^0 \times 5^3$$
$$2 = 2^1 \times 5^0 \;\text{ and } 5 = 2^0 \times 5^1.$$

These numbers are all multiples of the prime factors 2 and 5 which are factors of the base number 10.

We do not propose to prove this rule but it can be accepted that, for the scale of 10, any fraction of the form $\dfrac{p}{2^m \times 5^n}$ (m and n are of course whole numbers) gives a decimal with a finite number of places ; so that even a fraction with the tremendous denominator $2^{27} \times 5^{13}$ could eventually be worked out exactly, whatever the numerator might be. In other number scales there are similar rules. In the scale of 2, fractions with denominators of the form 2^m will give finite " decimal " fractions ; in the scales of 6 and 12 the corresponding denominators will be $2^m \times 3^n$—and so on. In very general terms we can summarise these rules and say : the fraction $\frac{1}{q}$ can be expressed as a finite series in the scale of b, *i.e.* as

$$\sum_{r=1}^{n} a_r b^{-r}$$

where n is finite, if q has no prime factors which are not also factors of b.

If in the fraction $\frac{p}{q}$ the denominator is a number which has one or more factors which are prime to the base number b, we find that an exact division is not possible. The result in this case does not end after a finite number of places ; the division never comes to an end but curiously enough it is possible to write down a complete answer. The reason is that after a certain number of places have been worked out there is a continual recurrence of a group of numbers. Such a decimal fraction is called a " *periodic recurring decimal.*" Here are some examples :—

$$\frac{1}{7} = 0 \cdot 142857142857142857 \ldots = 0 \cdot \dot{1}4285\dot{7}$$
$$\frac{3}{7} = 0 \cdot 428571428571 \ldots = 0 \cdot \dot{4}2857\dot{1}$$
$$\frac{4}{11} = 0 \cdot 363636 \ldots = 0 \cdot \dot{3}\dot{6}$$
$$\frac{2}{3} = 0 \cdot 6666 \ldots = 0 \cdot \dot{6}$$

When we get a decimal of this kind we indicate the recurring group of numbers by putting a dot over the first and last figures of the group.

The denominators of the fractions in the last examples consisted of one factor only. If the denominator contains a factor of 2 or of 5 (we are still only working in the scale of 10), then the recurring group is preceded by a group of figures which do not recur. For example $\frac{5}{6} = 0 \cdot 83333 \ldots$, $\frac{3}{26} = 0 \cdot 1153846153846 \ldots$; $\frac{5}{22} = 0 \cdot 2272727 \ldots$; $\frac{2929}{11100} = 0 \cdot 26387$.

We can summarise our knowledge of fractions expressed as decimals by saying that they are of three types :—

(1) Finite.
(2) Infinite with an infinitely recurring group only.
(3) Infinite with an infinitely recurring group preceded by a finite group.

These three types of decimal fractions cover all possible numbers like $\frac{p}{q}$ whatever factors the numbers p and q may or may not have. Chapter XVI introduced us to another kind of decimal, one in which the figures go on in an unending sequence with no trace of recurrence or any other kind of system. No fra tion can possibly give this kind of decimal, so our contention that irrational numbers like π, e, $\sqrt{2}$ and $\sqrt[3]{3}$ can never be exactly represented by a number like $\frac{p}{q}$ is confirmed. If p and q are both rational numbers the result of dividing p by q and expressing the result as a decimal is a rational number.

If it is possible to turn common fractions into periodic recurring decimals, then the reverse process, that of turning such decimals into common fractions, should also be possible. This reverse process will provide a test of our conclusions concerning decimals. Unfortunately, however, we do not yet know how to convert periodic decimals, even in the simple case of 0·3333. . . . Finite fractions are easy enough to deal with. For example, with 0·225 all we have to do is to write $\frac{225}{1000}$, that is, we divide it by 10 raised to the power whose index equals the number of decimal places. The fraction is of course normally reduced to its lowest form by cancelling common factors so that $\frac{225}{1000}$ becomes $\frac{9}{40}$.

The general statement of this rule would be

$$\sigma_m = \frac{\sum\limits_{r=1}^{m} a_r b^{m-r}}{b^m}.$$

In this forbidding looking formula there are two new terms to be explained. The Greek letter σ, sigma (not to be confused with the capital sigma Σ), is used to stand for a decimal with no figures in front of the decimal point. The suffix m is attached to σ to indicate that the decimal has m places. On the right-hand side of the formula b is as usual the base of the number scale and a_1, a_2 . . . a_r are the figures in the first, second . . . and nth decimal places. In the case of the example we have just done $m = 3$, $a_1 = 2$, $a_2 = 2$ and $a_3 = 5$, with $b = 10$. Therefore

$$\sigma_3 = \frac{2 \times 10^{3-1} + 2 \times 10^{3-2} + 5 \times 10^{3-3}}{10^3}$$

$$= \frac{200 + 20 + 5}{1000}$$

$$= \tfrac{225}{1000} \text{ as before!}$$

This formula looks rather clumsy when applied to the fraction we chose, but it has the advantage that it holds for any number system with place values.

For converting *purely* periodic decimals into common fractions we use the formula

$$\sigma_p = \frac{\sum\limits_{r=1}^{p} a_r b^{p-r}}{b^p - 1}.$$

Here p is the number of figures in the recurring period including any noughts. This formula too is not as formidable as it looks, though the proof of it is too long to include. In the scale of ten the denominator is always made up of nines because $(10^1 - 1)$ is 9, $(10^2 - 1)$ is 99, $(10^3 - 1)$ is 999, etc. If we want to turn $0 \cdot \dot{3}$ into a common fraction all we do is to write $\frac{3}{9}$ or $\frac{1}{3}$; similarly for $0 \cdot \dot{6}$ we write $\frac{6}{9}$ or $\frac{2}{3}$. Other examples are :—

$$0 \cdot \dot{3}\dot{6} \quad = \tfrac{3\,6}{9\,9} = \tfrac{4}{1\,1}$$
$$0 \cdot \dot{0}3\dot{7} \quad = \tfrac{0\,3\,7}{9\,9\,9} = \tfrac{1}{2\,7}$$
$$0 \cdot \dot{0}69\dot{3} \quad = \tfrac{0\,6\,9\,3}{9\,9\,9\,9} = \tfrac{7}{1\,0\,1}$$
$$0 \cdot \dot{0}7692\dot{3} = \tfrac{0\,7\,6\,9\,2\,3}{9\,9\,9\,9\,9\,9} = \tfrac{1}{1\,3}.$$

Of course the noughts in front of the numerators are disregarded but they are useful ås a check. The number of 9's in the denominator should be equal to the number of figures in the numerator, but only when the 0's after the decimal point are included.

If the recurring part of the decimal is preceded by figures which do not recur then the formula we have just used no longer holds. For such a case there is a formula which looks even worse than the first but which is just as easy to apply in practical cases. Here it is for any decimal in the scale of b :—

$$\sigma_{(n,\,p)} = \frac{\sum\limits_{r=1}^{n+p} a_r b^{n+p-r} - \sum\limits_{r=1}^{n} a_r b^{n-r}}{b^n(b^p - 1)}.$$

In this formula n is the number of figures which precede the recurring group and p is, as before, the number of figures in the repeated group. The application of the formula will be illustrated by two examples :—

(i) In the decimal $0 \cdot 2272727 \ldots$, or $0 \cdot 2\dot{2}\dot{7}$, there is one figure, 2, before the repeated group 27, so that $n = 1$, $p = 2$. The formula becomes

$$\frac{\sum\limits_{r=1}^{3} a_r 10^{3-r} - \sum\limits_{r=1}^{1} a_r 10^{1-r}}{10^1(10^2 - 1)}$$

in which $a_1 = 2$, $a_2 = 2$, $a_3 = 7$; so that

$$0 \cdot 2\dot{2}\dot{7} = \frac{227 - 2}{10 \times 99} = \tfrac{225}{990} = \tfrac{5}{22}.$$

(ii) In the decimal $0 \cdot 23\dot{4}7\dot{1}$, $n = 2$, $p = 3$, and so

$$0 \cdot 23\dot{4}7\dot{1} = \frac{\sum\limits_{r=1}^{5} a_r 10^{5-r} - \sum\limits_{r=1}^{2} a_r 10^{2-r}}{10^2(10^3 - 1)}$$

$$= \frac{23471 - 23}{99900} = \tfrac{23448}{99900} = \tfrac{1954}{8325}.$$

After a brief examination of these two results it is possible to state a rule of thumb which gives the fraction for mixed decimals in the scale of ten : for the numerator write down the decimal number up to the end of the first period and then the decimal number up to the end of the non-recurring period ; treat these as whole numbers and subtract the second from the first. For the denominator write down a number beginning with one 9 for each figure in the recurring group and ending with one 0 for each figure in the non-recurring group.

We have now given rules for converting any finite or recurring periodic decimal into fractions, although the fractions given by the formulæ will often need some reduction to bring them to their lowest terms.

We have accomplished great things and can look back over a large part of the realm of numbers. We understand the nature of whole numbers, fractions and irrational numbers, not only according to their absolute values but also as positive and negative numbers. Neither have we confined ourselves to a system of numbers based on 10. Moreover we have concerned ourselves with general numbers, with constants and unknowns. As we also have some knowledge of equations, of the solution of those with one unknown and of the nature of Diophantine equations, we are now equipped to begin in the next big subject. This is one of the biggest and most important branches of mathematics, namely the Theory of Functions. We have been well prepared for the task before us. There is no clear dividing line between elementary and higher mathematics but we can say that when we begin the Theory of Functions we are certainly leaving the elementary stages behind. From now on mathematics becomes more exciting, more adventurous but not intrinsically any more difficult. Many readers will from now on be breaking fresh ground. There is no need to be frightened. If we have been able to follow the argument well enough up till now, the rest of this book will be no more difficult.

CHAPTER XVIII

ALGEBRAIC FUNCTIONS

SUPPOSE we have a simple Diophantine equation before us,

$$3x - y = -5.$$

If we solve this by Euler's method,

$$x = \frac{y-5}{3} \text{ and } \frac{y-5}{3} = n,$$

then the solution is

$$\left.\begin{array}{l} x = n \\ y = 3n + 5 \end{array}\right\}.$$

Now let us substitute in this solution the numbers 1, 2, 3, 4 ...
and tabulate the results :—

n	x	y	Check
1	1	8	$3 - 8 = -5$
2	2	11	$6 - 11 = -5$
3	3	14	$9 - 14 = -5$
4	4	17	$12 - 17 = -5$
5	5	20	$15 - 20 = -5$
		etc.	

It seems that we have found an infinity of positive whole-number solutions. In the same way we can obtain an infinity of negative whole-number solutions :—

n	x	y	Check
-1	-1	$+2$	$-3 - (+2) = -5$
-2	-2	-1	$-6 - (-1) = -5$
-3	-3	-4	$-9 - (-4 = -5$
		etc.	

As we can see from the table, x and y are both negative when n is equal to $-2, -3, -4$ . . . and so on.

Now let us rearrange the Diophantine equation we began with and put it in the form

$$y = 3x + 5.$$

We are going to solve it again from this form but this time we will admit fractional solutions. All that we need to do is to substitute a fraction for x, say $\frac{1}{2}$, $\frac{1}{3}$, $\frac{2}{7}$ . . . and then find the corresponding value of y.

x	y
$\frac{1}{2}$	$6\frac{1}{2}$
$\frac{1}{3}$	6
$\frac{2}{7}$	$5\frac{6}{7}$
etc.	

If x is a fraction, y will almost always be a fraction too ; there is an example of the exceptional case in our table—when x is $\frac{1}{3}$ then y is 6. Now we know that there is an infinity of fractions between any two successive whole numbers, for example, between 2 and 3. If we let x take a fractional value lying between the numbers 2 and 3 there will be a corresponding value of y ; for the infinity of possible values of x there will be an infinity of corresponding values of y. If x can take any fractional value whatever, negative as well as positive, there will be a manifold infinity of fractional solutions.

Now that our knowledge of numbers goes beyond whole numbers and fractions we might ask ourselves what happens if we substitute an irrational number for x. The corresponding value for y will be irrational also. If we represent x by a non-periodic recurring decimal then the value of y will be a decimal of the same kind, that is, it will be an irrational number. For example, if $x = \sqrt{5}$, then $y = 3 . \sqrt{5} + 5$, which we can simplify only by working out the square root of 5. (A number like $3 . \sqrt{5} + 5$, which is a mixture of rational and irrational numbers, is called a surd.)

To sum up then, there are infinities of solutions of all kinds to our equation

$$y = 3x + 5$$

if we do not impose restrictions on the types of numbers we use for the values. For each value of x there is a corresponding value of y. We have found an infinity of positive integral solutions, negative integral solutions, positive and negative fractional solutions, and irrational number solutions. The number of possible solutions is overwhelming.

To show more clearly what is going on, let us look at a mechanical device which, as it "functions," behaves in a manner similar to the equation we have been dealing with.

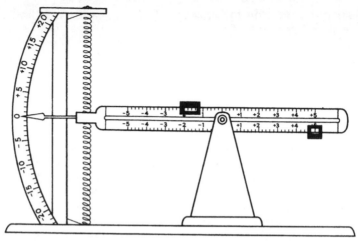

FIG. 13.

We will describe this device or instrument. It consists of a beam which is pivoted so that it can move up and down like the beam of a weighing machine. At one end it has a pointer which moves over a scale. The beam carries two sliding weights which can slide along the upper and lower edges of the beam. The distance of these weights from the pivot can be read from the scale of measurement along the beam.

There is an elementary law of mechanics which states that the effect of a weight in turning a beam is increased by moving the weight along the beam away from the pivot. A 5-lb. weight 1 inch from the pivot will balance a 1-lb. weight 5 inches from the pivot on the other side. The pointer of our weighing machine is at 0 on its scale when the beam is balanced. There is a spring attached to the pointer, fastened above and below it, which has the rather theoretical property of resisting a pull but offering no resistance to compression. Finally, suppose we make the weight which slides along the lower edge of the beam the constant of our equation and the weight on the upper edge the "x." The unit of weight will be the pound weight.

Now let us explain how the machine is operated. Here is the procedure : we select from our box of weights a 1-lb. weight and put it on the underside of the beam. This weight is going to represent the constant of the equation

$$y = 3x + 5,$$

so we clamp the weight at a point exactly 5 inches from the pivot. We clamp it in position because we know that for the purposes of this equation, in which the constant is 5, no change is possible. As soon as the weight is clamped in position the pointer will move upwards and come to rest on the scale mark 5. Now we have to deal with x. Glancing back at the equation, we notice that the coefficient of x is 3. So we select a 3-lb. weight from the box and put it on the upper edge of the beam. But where are we to let it rest ? The answer to this depends on the value of y which is given by the reading of the pointer on the pointer scale. Let us begin by putting the machine in a state of balance, when the pointer will read 0, that is $y = 0$. We shall find that this will happen when the 3-lb. weight is on the mark $-1\frac{2}{3}$, which means, when it is $1\frac{2}{3}$ inches on the left-hand side of the pivot. (All left-hand side markings are negative.) With this arrangement the two weights are balanced about the pivot because $3 \times 1\frac{2}{3} = 1 \times 5$.

In some ways it would be neater to write our equation in the form
$$y = 3x + 5$$
$$y = 3x^1 + 5x^0.$$

This does not affect our equation since x^0 is always equal to 1. To represent the term $3x$ on our machine we used a 3-lb. weight to stand for the coefficient 3 and a scale position to represent the unknown x. It would be consistent then to represent the constant $5x^0$ by a 5-lb. weight at 1 inch from the pivot. We shall come back to this later.

Since the " constant " weight is fixed, we have only one weight, the 3-lb. weight on the upper edge of the beam, to play with. This can be moved between the marks $+5$ to -5 inches ; that is, the value of x can range between $+5$ and -5. For any position of the 3-lb. weight there will be a corresponding reading of the pointer, i.e., for every value of x there will be a corresponding value of y. If x is at $+2$ inches then the pointer reading will be $+11$ since $3 \times 2 + 5 = 11$.

So, if we vary x at will, y is forced to vary on our machine. This relationship is the basis of the idea of " *Function*." If two quantities, x and y, are connected so that when x varies y is forced to vary also, then y is said to be a function of x. Our machine has demonstrated the connection between x and y in the case of the equation $y = 3x + 5$. (Of course it only

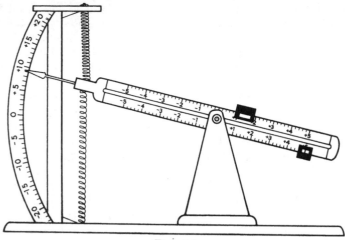

FIG. 14.

did so because we constructed the machine in order to obey the " law " expressed by the equation.)

We shall as usual need a shorthand form for expressing this new idea. We write

$$y = f(x)$$

and read this as " y equals function x " or " y is a function of x," meaning that y is in some unstated but systematic way related to x.

It is usual to call x and y " *variables* " and to say that x is the " independent " and y the " dependent " variable. These terms are perhaps a little formal and colourless for a book which is intended to be an introduction to mathematics ; we will rename the variables and call x " *free* " and y " *forced.*"

Having agreed on the words we are going to use and how we are going to use them, we can take the first steps in the theory of functions. The simplest way to begin will be to use our machine again. Suppose we move the 3-lb. weight very slightly away from the pivot, then the pointer will move up, of course only very slightly.

We can imagine that the upper edge of the beam represents that part of the number line marked off between + 5 and — 5. The edge is a continuous line and so is the number line. As we push the 3-lb. weight from one mark to another on this edge we can imagine the value of x passing through all the values that lie between these two marks on the number line. But what lies between these two number line marks ? All the common fractions and irrational numbers that have their place between the two marks. So that when we move the 3-lb. weight continuously along the edge, we are letting x take every possible value in turn of whole numbers, fractions and irrational numbers.

This idea of " continuity," of x taking every possible value between two marks on the number line, is of the highest importance in the theory of functions. It is due in the main to the mathematician Weierstrass. We will rest content for the present with this mere mention of continuity. When we return to it in a later chapter we will clarify the idea, using geometrical illustrations.

We are now going to use our very limited knowledge of functions in a task whose meaning and purpose will not at first be apparent. As it requires only a simple operation in algebra there is no reason why we should defer it.

The question we ask is : " What happens to the y pointer when x grows by a definite small amount at any place on the line ? " Presumably the y pointer also moves by a certain amount. We can use the relationship of the increase of x and that of y as a measure of their variability. Furthermore, if we give this increase in x quite a general form and let it occur at any place on the line, it is obvious that the pointer will come to rest at a correspondingly increased position of y.

So this general x which has no definite value is increased by a certain finite amount which we will call Δx. The Δ we have just used is the Greek capital letter " delta." Placed before the x (Δx) it is read as " delta x," and means " a small increase of x." Now, because of the law which connects x and y, we expect the " forced " variable y to be changed when the " free " variable x is changed. Looking back at the machine which is still set up to represent the equation $y = 3x + 5$, the 1-lb. weight is still fixed on the lower edge at the $+ 5$ mark and the 3-lb. weight is just " somewhere " on the upper edge.

In algebraic terms the question now becomes : " What is the relation between Δy and Δx for this particular function ? What is the ' forced ' Δy for Δx, Δx being the change in the ' free ' variable x ? "

Without more ado let us tackle the problem purely mathematically. If we want to incorporate the increments (the

small increases) in our equation (or function) we must write
the equation thus :—

$$y + \Delta y = 3(x + \Delta x) + 5$$

for, as the result of the increase in x (or the moving of the
weight), x has become $x + \Delta x$ and it follows that y must
forcibly become $y + \Delta y$.

So far then we do not know how big x is, nor do we know
how big Δx is. We only know that Δx is finite. It could be
as big as we like but it will be convenient to make it small.
So, if

$$y + \Delta y = 3(x + \Delta x) + 5,$$

what we want to discover is the relationship of Δy to Δx, that
is $\Delta y : \Delta x$ or the ratio of Δy to Δx. Let us multiply in order
to get rid of the bracket :—

$$y + \Delta y = 3x + 3\Delta x + 5.$$

At this point we conveniently remember that $y = 3x + 5$
and that we are justified in taking y from the left-hand side
and $3x + 5$ from the right, just as we were able to remove
apples and ounce weights from the scale pans in Chapter XIII.
So

$$y + \Delta y = 3x + 3\Delta x + 5$$

subtracting $\qquad y = 3x + 5$

$$\Delta y = 3\Delta x$$

from which it follows that $\dfrac{\Delta y}{\Delta x} = 3$.

This is the answer to the question we set ourselves. We
know furthermore that the result, the ratio that is, does not
in any way depend upon the size of Δx. In fact we can make
it as small as we like, almost up to the very limit of 0. We
are going to make Δx so small that we find the value of $\dfrac{\Delta y}{\Delta x}$
when, in Newton's phrase, the increment " is about to dis-
appear " into 0. It is at this " disappearing point " that we
are particularly interested in the ratio of Δy to Δx and we
find the ratio of Δy to Δx at this point. We then allow
$\Delta x \rightarrow 0$ and we write the limiting value (see Chapter XII) of
the ratio $\dfrac{\Delta y}{\Delta x}$ as $\dfrac{dy}{dx}$.

So that finally $\dfrac{dy}{dx} = \underset{\Delta x \to 0}{L}t\ \dfrac{\Delta y}{\Delta x} = 3.$

Now we must confess that we have found a " differential coefficient." We have found that the differential coefficient of the function $y = 3x + 5$ has the value 3. Sometimes $\dfrac{dy}{dx}$ is written as y' to indicate that it is the " first " differential coefficient of a function $y = f(x)$.

The first thing to notice is that the differential coefficient which we have just found is the same whatever x may be. Whatever value of x we begin with, the increase of x by the smallest possible amount Δx always gives the same ratio for the corresponding increase of y to $y + \Delta y$, or $\Delta y : \Delta x = 3 : 1$ for all values of x.

The second thing to notice is that the constant has nothing to do with the result. In fact, if it had been omitted we should have had

$$y = 3x$$
$$y + \Delta y = 3(x + \Delta x)$$
$$y + \Delta y = 3x + 3\Delta x$$

Subtracting
$$y \doteq 3x$$
_____
$$\Delta y = 3\Delta x$$

therefore $\qquad \dfrac{\Delta y}{\Delta x} = 3$ and $\dfrac{dy}{dx} = 3.$

Why the constant has no effect on the coefficient we shall explain later. It can be seen from the working of our machine that the value of the constant, *i.e.*, the clamped weight, has no effect on the forced change in the value of y when there is a change in the value of x. The differential coefficient shows us how the function varies for all possible values of x.

We will now try our little knowledge of the differential calculus on something new, the " quadratic function,"

$$y = 2x^2 + 7.$$

There is no possibility of using our machine any more so we will throw it away and put our trust in mathematics. According to our method the first step is to write

$$y + \Delta y = 2(x + \Delta x)^2 + 7.$$

Since we have not yet discussed how to deal with $(x + \Delta x)^2$ we will flatly state that

$$(x + \Delta x)^2 = (x + \Delta x)(x + \Delta x)$$
$$= x^2 + x\Delta x + x\Delta x + (\Delta x)^2$$
$$= x^2 + 2x\Delta x + (\Delta x)^2$$

Now we perform the usual trick

$$y + \Delta y = 2x^2 + 4x\Delta x + 2(\Delta x)^2 + 7$$

Subtracting $\qquad y = 2x^2 + 7$

$$\Delta y = 4x\Delta x + 2(\Delta x)^2$$

Our first need is to find the ratio of $\dfrac{\Delta y}{\Delta x}$ and then secondly to find the limiting value of this ratio when $\Delta x \to 0$. Returning to our last equation we divide by Δx ; $\dfrac{\Delta y}{\Delta x} = 4x + 2\Delta x$.

This equation contains a term which did not arise in the calculation of the other differential coefficient we have done, *i.e.* the term $2\Delta x$. It does not matter. Our next step is to find the limit of the left-hand side of this equation when $\Delta x \to 0$. What is the limit of $2\Delta x$? Clearly, $2 \times 0 = 0$. Therefore, finally,

$$\frac{dy}{dx} = \mathop{L}_{\Delta x \to 0} t \frac{\Delta y}{\Delta x}$$
$$= \mathop{L}_{\Delta x \to 0} t \, (4x + 2\Delta x)$$
$$= 4x.$$

The result for the quadratic function shows a new feature. In the last result we obtained the number 3. Now we have the quantity $4x$, that is, a quantity which depends on the value chosen for x.

It would be poor psychology to pursue the methods of the differential calculus any further in this abstract kind of way. Historically it developed by making use of geometry and we will reconsider it later in this way also. This will have the advantage of enabling the reader to visualise the differential coefficient and we can usually understand more easily what we can see.

CHAPTER XIX

THE THEOREM OF PYTHAGORAS

WE will go back in history to the time of the ancient Egyptian and Hindu civilisations. Those who know the buildings of ancient Egypt which are still to be seen have always been impressed by the massiveness and precision of the architecture. The builders owed much to the knowledge and accuracy of the *harpedonaptœ* or "rope-stretchers," who did the work of surveying. They had practical methods for constructing angles, particularly the right angle. If a large rectangular temple was to be built it was obviously important that the builders should be able to construct right angles very accurately. Now the rope-stretchers, who formed a guild belonging to the priesthood, began to carry out their geometrical ceremonies when the foundation stone was being laid. Their most important instrument was a rope divided by knots into 3 parts in the ratio 5 : 3 : 4.

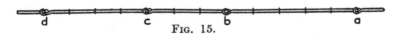

d   c   FIG. 15.   b   a

Let us call the knots a, b, c and d. To make a right angle at c the part of the rope between b and c is pegged at b and c. Then the part c to d is moved round to make approximately a right angle and the part b to a is also moved round till a and d coincide and the ropes are taut. When the ropes are pegged in this position the angle at c is exactly a right angle.

This triangle, which has its sides in the ratio 3 : 4 : 5, is generally known as the " Egyptian triangle." That it is at least approximately a right-angled triangle can be seen from the figure.

Not only the Egyptian but also the ancient Hindu priesthood had a similar method of marking out right angles. Rather surprisingly they had another triangle whose sides were in a ratio 5 : 12 : 13.

We must remind ourselves of some facts about the geometry of the triangle. A right angle contains 90° (degrees) and the

129

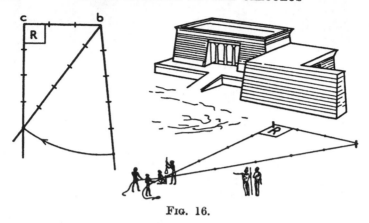

FIG. 16.

sum of the three angles of any triangle is always 180°, or two right angles. Moreover, the longest side is opposite the largest angle and the shortest side opposite the smallest angle. In the right-angled triangle therefore the largest side, or the "hypotenuse," as it is called, is opposite the right angle. In the right-angled triangle the right angle is equal to the sum of the other two angles. We might reasonably expect to find some simple relation also between the hypotenuse and the other two sides of the triangle. It is obvious that the relation we expect to find is not merely that the hypotenuse is equal to the sum of the other two sides, for in the Egyptian triangle $4 + 3$ is not equal to 5. If the numbers are raised to the second power, however, we find a possible relationship :—

$$3^2 + 4^2 = 9 + 16 = 25 = 5^2$$
$$5^2 + 12^2 = 25 + 144 = 169 = 13^2$$

This simple relationship of the squared sides is one of the most important theorems of geometry, the theorem of Pythagoras. In the Middle Ages scholars called it the "pons asinorum," or "bridge of asses," because an understanding of the theorem was a test which divided the good students from the bad.

The general statement of this theorem is : if the hypotenuse of a right-angled triangle is called c and the other two sides are a and b, then $$a^2 + b^2 = c^2.$$

There are many proofs of the theorem. We will give an obvious demonstration of its truth.

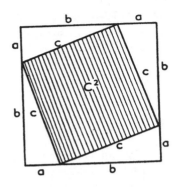 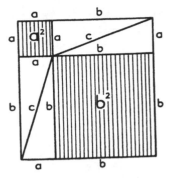

FIG. 17.

The first square is made up of the shaded square and four triangles. Its area is therefore $c^2 + 4\Delta$, using Δ to represent the area of one of the triangles which has the sides a, b, c.

The second square is made up of the two shaded squares and the four triangles. Its area is therefore $a^2 + b^2 + 4\Delta$.

But since the big squares are both made up of sides which are $(a + b)$ long, they are equal in area. Therefore

$$a^2 + b^2 + 4\Delta = c^2 + 4\Delta.$$

Removing 4Δ from both sides

$$a^2 + b^2 = c^2.$$

Whatever lengths we choose for a, b and c we shall find that this theorem is true for all right-angled triangles.

We are going to use the formula we have just derived as an equation, that is to say we are going to use it to find one side of a right-angled triangle when we know the other two. It might be helpful to rewrite the equation in all possible forms :—

$$c^2 = a^2 + b^2$$
$$a^2 = c^2 - b^2$$
$$b^2 = c^2 - a^2.$$

To find the length of a side from these equations we shall have to take the square root of both sides of the equations :—

$$c = \sqrt{a^2 + b^2}$$
$$a = \sqrt{c^2 - b^2}$$
$$b = \sqrt{c^2 - a^2}.$$

But we already know that many square roots give us irrational numbers. Let us look at an instructive example.

We will make the two shorter sides of a right-angled triangle equal to each other so that we can write $b = a$ in our formulæ. For such an " isosceles " right-angled triangle as it is called, we write the equation

$$c^2 = a^2 + a^2$$
$$= 2a^2$$
$$c = \sqrt{2a^2} = a\sqrt{2},$$

since the square root of a^2 is a.

As 2 is not a perfect square (that is, it is not the square of a whole number), the square root of 2 must be irrational and therefore $a\sqrt{2}$ is irrational also. Incidentally, by putting two equal isosceles right-angled triangles together we get a square with sides of length a and diagonal of length c (Fig. 18).

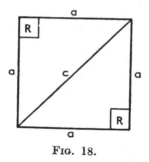

FIG. 18.

The ratio of a diagonal to the side of a square is therefore irrational, and vice versa. As $c = a\sqrt{2}$, then $a = \dfrac{c}{\sqrt{2}}$. If a is rational, then c is irrational ; if c is rational, then a is irrational.

Irrationality in geometry therefore is a matter of the proportion of two lengths. If two lengths when compared give a ratio which is an irrational number, then they are said to be " incommensurable." Pythagoras gave symbolic meaning to his system of numbers and incommensurability was the symbol for all things living because life is not " measurable."

We do not want to become involved with symbolism however. We need a rule which will give us all possible whole-number right-angled triangles, not only the Egyptian and Hindu ones, but as many others as we care to find.

Here is the rule : if u and v are two positive whole numbers, $u > v$, then the formulæ $c = u^2 + v^2$, $a = u^2 - v^2$, $b = 2uv$ gives the sides of rational right-angled triangles.

u	v	$u^2 + v^2 = c$	$u^2 - v^2 = a$	$2uv = b$
2	1	5	3	4
3	1	10	8	6
3	2	13	5	12
4	1	17	15	8
4	2	20	12	16
4	3	25	7	24
5	1	26	24	10
5	2	29	21	20

etc.

The table we have just set out gives whole number solutions. We can of course multiply any solution by any number, whole, fractional or irrational, and still get sides of true right-angled triangles., For instance, the Egyptian triangle of sides 5, 4, 3, gives rise to a series of triangles such as 10, 8, 6 ; 80, 64, 48 ; or $\frac{5}{4}$, 1, $\frac{3}{4}$. Generally, if, a b and c is a whole-number solution, then ma, mb and mc will also be a solution when m is any rational number, integral or fractional.

CHAPTER XX

FUNCTIONS OF ANGLES

LET us stick to our task which at this stage demands our whole attention. When we propounded the theorem of Pythagoras we indicated that there was a necessary relationship between the sides and the angles of a right-angled triangle. The special branch of mathematics which deals with this relationship is called trigonometry (from the Greek *tri* = three, *gonio* = angle, *metri* = measurement). We shall not linger over the consideration of this special subject but we ought to make the acquaintance of some of its basic ideas as they will be needed later in the development of the differential calculus.

We will draw a circle of any size we fancy and divide it into four quarters or " quadrants " by drawing two diameters through the circle at right angles to each other (see Fig. 19).

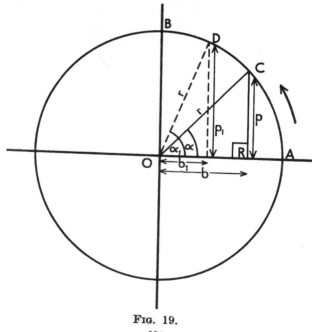

FIG. 19.

134

If we imagine that the radius r is a movable arm and that it moves from OA in the direction of the arrow to OB, then OA and the moving area will have formed all angles between $0°$ and $90°$. We will assume that it is known that the whole circle contains $360°$ and therefore a quadrant contains $90°$.

If the moving arm comes to rest at a position OC between OA and OB, the angle between OC and OA must be some angle between $0°$ and $90°$; we do not know, nor does it matter, what this angle is exactly. We will call it α (alpha). Now we will draw from C a line at right angles to OA. This makes a right-angled triangle with its hypotenuse r, base b and height p. In two cases the right-angled triangle will disappear and become a straight line. Firstly this will happen when the moving arm coincides with OA and secondly when it coincides with OB. In between these extremes all possible right-angled triangles can occur, each of course with a different angle α.

The chief problem of trigonometry is how the size of the angle α is to be determined if we know only the lengths of the sides of the right-angled triangle. It is obvious that there is some relationship, for, if another triangle is drawn (shown in Fig. 19 by dotted lines), whose angle α is greater than $45°$; the figure shows that the new sides p_1 and b_1 are different from p and b though the hypotenuse r remains the same.

If the hypotenuse were always the same as it is in Fig. 19, the determination of the size of the angle α would be simple. All we should need to do would be to construct a table showing the angle α which corresponded to any given height or base of a triangle. In actual fact the hypotenuse can be of any length, so we need to know the ratio of two of the sides of a triangle—any two sides will do. Casting back our minds to the chapter on Combination (Chapter VII), we can see that the three sides of a triangle give six possible ratios : $\dfrac{r}{p}$, $\dfrac{r}{b}$, $\dfrac{p}{b}$, $\dfrac{p}{r}$, $\dfrac{b}{r}$, $\dfrac{b}{p}$. These ratios are the simple functions of an angle.

Let us draw the right-angled triangle again :—

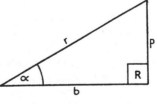

Fig. 20.

Now we will make a list of the names of the 6 functions :—

$\frac{p}{r}$ is the " Sine "   (Opposite $\div$ Hypotenuse).

$\frac{b}{r}$ is the " Cosine "   (Adjacent $\div$ Hypotenuse).

$\frac{p}{b}$ is the " Tangent "   (Opposite $\div$ Adjacent).

$\frac{b}{p}$ is the " Cotangent " (Adjacent $\div$ Opposite).

$\frac{r}{b}$ is the " Secant "   (Hypotenuse $\div$ Adjacent).

$\frac{r}{p}$ is the " Cosecant "   (Hypotenuse $\div$ Opposite).

Normally we need only the first four of these functions, and for our immediate purpose we need only one of these, that is the " tangent." It would be somewhat high handed to dismiss them so summarily so we shall look at one of them more closely to see what they are like. The " sine," for example, of the angle α is the ratio of the opposite side to the hypotenuse, $\frac{p}{r}$. We can simplify this ratio by making r equal to 1.

There is no reason why we should not, because the units in which the sides are measured can be any we choose. So we make the circle a " unit circle," that is a circle with its radius equal to 1 unit, whatever that unit may be. By introducing this dodge our " sine " becomes $\frac{p}{1}$ or just p and we have managed to measure an angle by transforming it into a length. So long as we measure the height or the opposite side of the triangle as a fraction of the radius, we shall find the value of the sine of the angle α directly.

In Fig. 21 therefore, the lengths p, p_1, p_2, p_3, etc., are the measures of the sines of the angles α, α_1, α_2, α_3, etc. Furthermore, the sine of $0°$ is obviously 0 ; the sine of $90°$, when the length p is equal to the radius, is 1. In this quadrant the sine of the angle increases as the angle itself increases from $0°$ to $90°$ and it takes every possible number value (including irrational numbers) between 0 and 1. That the sine of an angle is not

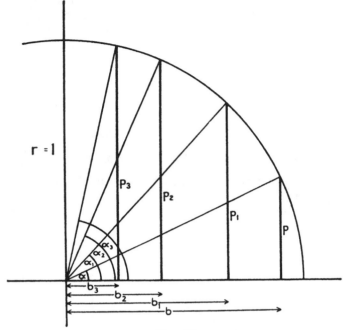

Fig. 21.

always a rational number is demonstrated when $\alpha = 45°$; for this angle p is half the diagonal of a square whose side is r. Using the theorem of Pythagoras,

$$r^2 = p^2 + p^2 = 2p^2$$

$$p^2 = \frac{r^2}{2}$$

or $\qquad p = \frac{r}{\sqrt{2}} \qquad = \frac{1}{\sqrt{2}}$ for our unit circle.

In other words, the sine of 45° is $\frac{1}{\sqrt{2}}$, which we can write as

$$\frac{1}{\sqrt{2}} \times \frac{\sqrt{2}}{\sqrt{2}} = \frac{\sqrt{2}}{2},$$

clearly an irrational number.

We will not pursue this line of thought further but will

remark that in practice the actual values of the sines and of the other ratios are rarely used. Any book of logarithms contains tables of the logarithms of these angle functions ; in four-figure tables most of the values are given correct to 1 minute (1 degree = 60 minutes, 1 minute = 60 seconds ; or, $1° = 60'$, $1' = 60''$).

Now let us examine the tangent function in like manner, for it is of special importance to us. It seems likely to have something to do with the idea of a tangent, that is a straight line touching a curve. We will now construct a machine which will show the connection. The tangent of α is $\frac{p}{b}$; we can simplify this ratio by making b the unit of length, $i.e.$ by putting b equal to 1. Therefore the moving arm (or " vector ") is no longer the radius itself but an extension of it beyond the circle.

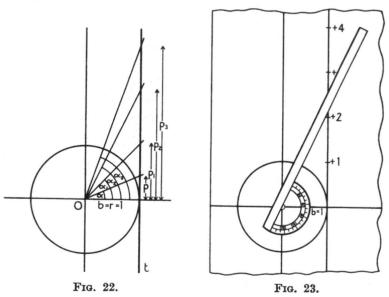

FIG. 22. FIG. 23.

We have now transformed our ratio $p : b$ into $p : 1$, i.e. into a section of a tangent to the unit circle. Now the base of the triangle is always equal to the radius of the unit circle, while the height and the hypotenuse can vary. The apparatus shown in Fig. 23 is a direct representation of this scheme.

We can see the unit circle, the radius = 1, the tangent (on which a scale in terms of the radius is marked), and lastly the side which rotates about the centre of the circle and slides along the tangent. The angle α can be read directly from a protractor fixed on the hypotenuse.

When the angle α is 0, the tangent of α is also 0 because $\frac{0}{1} = 0$. But the value of the tangent function grows rapidly. When α = 45°, tan α (the conventional way of writing " tangent of α ") is 1 ; tan 60° = 1·732 correct to four significant figures ; tan 70° = 2·748 ; tan 80° = 5·671 ; tan 85° = 11·43 ; tan 89° = 57·29; tan 89⅝° = 343·8. Finally, as we approach 90° the section of the tangent cut off increases in length without limit because at 90° the moving arm can no longer reach the tangent.

It may well be asked what the purpose of trigonometry is. We said that we needed the tangent and its function most urgently for our pursuit of higher mathematics ; but that cannot be the sole reason for building up such a complicated and therefore difficult branch of knowledge. Trigonometry is needed for the calculation of sides or angles whenever we are dealing with triangles. Astronomical measurements, navigation and surveying all are based on trigonometrical methods. The height of a mountain such as Mount Everest can be measured exactly from a distance. The method is to select a measured base line and then to find the angles of elevation of the mountain peak from the two ends of the line, using the accurate measuring telescope called a theodolite. The height of the mountain then forms one side of a right-angled triangle whose other sides and angles can be calculated.

In Fig. 24 the angles α and β are the angles measured by

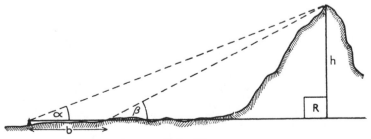

FIG. 24.

the theodolite and b is the base line whose length we know. By using the relatively simple trigonometry of the right-angled triangle, we can arrive at a formula which gives the height in terms of quantities we know. We will not prove this formula here but will state it in case the reader would like to work out an example. It is

$$h = \frac{b \cdot \sin \alpha \cdot \sin \beta}{\sin (\alpha - \beta)}.$$

A similar method is also used in gunnery when determining the position of a distant target.

Interesting as this subject of trigonometry is, it behoves us to turn our attention to a further extension of our idea of number which we need to carry us on to higher spheres of mathematics. This trigonometry of measurement which we have glanced at does not lead to any important step in mathematical theory, although it is of great practical value. It is more important for our purpose to return to the consideration of numbers and to examine yet another class of numbers, the so-called imaginary numbers.

CHAPTER XXI

IMAGINARY AND COMPLEX NUMBERS

ACCORDING to our usual custom our introduction to this difficult subject will be made by considering the simplest cases. First of all let us recall that we received our first mathematical shock when, in finding roots of numbers, we came upon irrational numbers. Calculation of roots leads us again to this new subject, *imaginary numbers*. In some respects they have been unfortunate in their name " imaginary." The word now conveys to us the sense of something mysterious, whereas, if we return to the sense of the Latin word from which it is derived, we shall see that it means " a representation, a counterpart." Earlier mathematicians, puzzled by the problem which we are now going to face, called these numbers " impossible."

How do imaginary numbers arise ? When we were learning how to deal with signs we found out how to simplify a given sum by manipulating the signs according to the rules. When however the last stage of simplification had been reached we should not have been able to tell merely by looking at the answer what the signs of the original numbers in the calculation had been. There is no means of telling whether a^2 is derived from $(+a) \times (+a)$ or from $(-a) \times (-a)$. Ordinarily this is of no great importance. If a^2 is known to have been derived from $(-a) \times (-a)$ we shall arrive at the same result in any further calculation with a^2, as we would with a^2 derived from $(+a) \times (+a)$. For example, when we divide a^2 by $(+a)$ we could have either

$$\frac{(+a) \times (+a)}{(+a)} = (+a)$$

or
$$\frac{(-a) \times (-a)}{(+a)} = \frac{(-a) \times [(+a) \times (-1)]}{(+a)}$$
$$= \frac{(-a) \times (+a) \times (-1)}{+a}$$
$$= (-a) \times (-1) = (+a).$$

The result is the same.

Similarly, if we divide a^2 (thinking of it as $(+a) \times (+a)$ or as $(-a) \times (-a)$) by $(-a)$, the result $(-a)$ will be obtained in both cases.

Where roots are concerned it is different; though $(+a) \times (+a)$ can only equal a^2 and $(-a) \times (-a)$ can also only equal a^2, the square root of a^2, or $\sqrt{a^2}$, has not a single value. Although we can be sure that the root will have the absolute value a (written $|a|$), we know nothing at all about its sign and can never know from the symbol $\sqrt{a^2}$ alone. We have to admit our ignorance of the sign and write $\sqrt{a^2} = \pm a$, that is either $+a$ or $-a$. But this doubt about the sign does not arise in every case of finding roots. Take $\sqrt[3]{a^3}$ for example; there is only one root and that is $(+a)$, for $(-a) \times (-a) \times (-a)$ is $(-a)^3$. It therefore follows that $\sqrt[3]{-a^3}$ is $(-a)$. When we come to the fourth root we again find there is a doubt about the sign for both $(+a) \times (+a) \times (+a) \times (+a)$ and $(-a) \times (-a) \times (-a) \times (-a)$ are equal to $(+a^4)$. Hence $\sqrt[4]{a^4}$ is $\pm a$. The pattern is becoming clear; when we take an even root, there is a double-valued answer; when we take an odd root, there is a single-valued answer. We can write this in a general way. Suppose that an " nth " root of r is s, then if n is even,

$$\sqrt[n]{r} = \pm s$$

if n is odd

$$\sqrt[n]{r} = +s$$

$$\sqrt[n]{-r} = -s.$$

So far so good. But suppose we were asked to find the $\sqrt[n]{-r}$ when n is an even number; we should not be able to give an answer. In the realm of numbers, as we have investigated it so far, we have not found any number which could satisfy the demand we now make. Any number raised to an even power must be positive. Yet here we are asked to find a number which, when raised to an even power, is negative. There is no such number, be it general, concrete, integral, fractional, irrational, positive or negative.

It seems therefore that if an answer to the question is to be given a new kind of number will have to come to our rescue. It must have the property of giving a negative number when raised to an even power. Let us begin with the simplest possible case when r is 1 and n is 2, $\sqrt[2]{-1}$, the square root of (-1). This is one of our new numbers and one which mathematicians have found so useful that they have given it a special symbol, namely i. Its usefulness can be seen at once ; if we have to find the $\sqrt{-15}$ we write

$$\sqrt{-15} = \sqrt{(-1) \times 15}$$
$$= \sqrt{(-1)} \times \sqrt{15}$$
$$= i\sqrt{15}$$

Let us play with our new number and see what happens when we use it. Suppose we try to find $\sqrt{(-9) . (-4)}$. We could manage without using our new number at all because

$$\sqrt{(-9) \times (-4)} = \sqrt{36} = \pm 6.$$

However, if we do use the number i, we could write

$$\sqrt{(-9) \times (-4)} = \sqrt{(-9)} \times \sqrt{(-4)} = i\sqrt{9} \times i\sqrt{4}$$
$$= i^2 \times \sqrt{36}$$

and since $i^2 = -1$ this equals $(-1) \times (\pm 6) = \mp 6$.

The two results may look different because the signs $\pm$ and $\mp$ are reversed, but all we are saying is that in the first case the answer is $+6$ or -6, and in the second case the answer is -6 or $+6$. There is no serious difference ; $+6$ and -6 are the two possible answers. It does not matter in which order they come. There are two answers and we can choose which we like.

A much more complicated example showing how the number i is used arises from a story about the famous Dutch physicist Huyghens and the German mathematician Leibniz. Leibniz sent Huyghens this problem : " Show that

$$\sqrt{1 + \sqrt{-3}} + \sqrt{1 - \sqrt{-3}} = 2 \cdot 449897 \ldots = \sqrt{6}."$$

This unpleasant-looking command is not difficult to perform nor is its surprising result difficult to verify if we use two important rules of thumb in algebra :—

(1) $(a + b)(a - b) = a^2 - ab + ab - b^2 = a^2 - b^2$, that is, the product of the sum and the difference of two numbers is equal to the difference of the squares of the two numbers.

(2) $(a + b)(a + b) = a^2 + ab + ab + b^2 = a^2 + 2ab + b^2$, that is, the square of the sum of two numbers equals the sum of the squares of the two numbers together with twice their product.

Instead of proving directly that $\sqrt{1 + \sqrt{-3}} + \sqrt{1 - \sqrt{-3}} = 2 \cdot 4494897 \ldots$ we shall prove that its square is equal to 6. So, writing i for $\sqrt{-1}$ we have

$$(\sqrt{1 + i\sqrt{3}} + \sqrt{1 - i\sqrt{3}})^2$$
$$= 1 + i\sqrt{3} + 2\sqrt{1 + i\sqrt{3}} \cdot \sqrt{1 - i\sqrt{3}} + 1 - i\sqrt{3}$$

(using the second rule)

$$= 2 + 2\sqrt{(1 + i\sqrt{3})(1 - i\sqrt{3})}$$
$$= 2 + 2\sqrt{1 - i^2 \cdot 3}$$

(using the first rule)

$$= 2 + 2\sqrt{1 - (-1) \times 3}$$
$$= 2 + 2\sqrt{1 + 3}$$
$$= 2 + 2\sqrt{4}$$
$$= 2 + 2 \times 2$$
$$= 2 + 4 = 6$$

Leibniz's contention is verified.

In this last example we used the expressions $(1 + i\sqrt{3})$ and $(1 - i\sqrt{3})$; numbers of this kind have special names and important properties. If a and b are two real numbers, then any number of the form $(a + ib)$ is called a " *complex number* " and the two complex numbers given by $(a + ib)$ and $(a - ib)$ are said to form a " *conjugate pair.*" Using the algebraic rules (better called " identities ") again, we can show some of the most important properties of conjugate pairs. The multiplication of two conjugate complex numbers gives a real number, since (using the first identity)

$$(a + ib)(a - ib) = a^2 - i^2 \cdot b^2 = a^2 - (-1) \cdot b^2$$
$$= a^2 + b^2.$$

The sum of the squares of two complex conjugate numbers also gives a real number as the imaginary parts cancel out. Using the second identity,

$$(a + ib)^2 + (a - ib)^2 = a^2 + 2iab + i^2b^2 + a^2 - 2iab + i^2b^2$$
$$= 2a^2 + 2i^2b^2$$
$$= 2a^2 - 2b^2.$$

Before we go further in pursuit of imaginary numbers we will make a table showing all the types of number we have now met.

1. Real Numbers.
    A. Rational Numbers,
        (*a*) Integers (2, 4, 99).
        (*b*) Fractions ($\frac{5}{7}$, 0·25, 0·$\dot{3}$).
    B. Irrational Numbers ($\sqrt{5}$, 3·14159 . . . $= \pi$, etc.).
        (*a*) Surds (5 $+$ $\sqrt{5}$).
2. Imaginary Numbers.
        (*a*) The number $i = \sqrt{-1}$.
        (*b*) Complex numbers, $a + ib$.
        (*c*) Conjugate complex numbers $(a + ib)$, $(a - ib)$.
We can tabulate numbers in another way :—
  (*a*) Concrete numbers (7, $\sqrt{5}$, 5 $+$ $i\sqrt{3}$, etc.).

  (*b*) Generalised numbers (a, c, p^n, $\dfrac{r}{3}$, $a - ib$, etc.).

  (*c*) Unknown numbers (x, y, z, etc.).
  (*d*) Variable numbers ($y = f(x)$, $z = 5y + 3x$, etc.).
Yet another way of tabulating would be
A. Positive and negative numbers ($+$ 5, $-$ $\sqrt{a}$, $\pm$ $3x$, etc.).
B. Absolute numbers ($\mid 7 \mid$, $\mid \sqrt{a} \mid$, $\mid a + ib \mid$, etc.).

We have not covered the whole field of numbers but we have sufficient knowledge of them to be able to delve more deeply into mathematics. There are other types of numbers of which we may mention Hamilton's quaternions and transcendental numbers. We can be satisfied however with our present achievements.

Now let us go back to the " number-line " which we have already found helpful in visualising our earlier number systems. Let us see if there is a place on the number-line for our new and somewhat elusive imaginary numbers. We will draw the line and see what we can do about it.

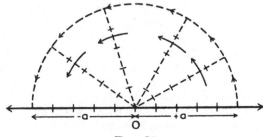

Fig. 25.

We want to be able to work in general terms, so we will take any section of the positive part of the number-line and call it " a." In our diagram (Fig. 25) we have chosen the number 5 to be " a." We could of course have chosen any other number. Let us imagine the line to be hinged at 0, the zero. We can now turn the line about 0 and if we turn it far enough we can make it coincide with the negative arm so that the original $(+a)$ fits exactly over the point $(-a)$. As a result of this manipulation we have established two conventions. First, we have made the left-hand side of the number-line the part along which we deploy the minus quantities. Secondly, we have decided to make the anti-clockwise rotation of the arm about 0 the *positive* direction of rotation. How was this transformation of $(+a)$ into $(-a)$ achieved? Geometrically we have turned the section of the line $(+a)$ through half a circle or 180° about the zero point. If we want to give it an arithmetical interpretation we find that the only one which fits into our scheme is that the rotation through 180° is equivalent to multiplying by (-1). Though this interpretation may not look very promising, it is in fact a very fruitful idea. We know well that $(+a) \times (-1) = (-a)$. The (-1) we will call the "turning factor." If we wish to test our scheme further, turning $(-a)$ through 180° should give us $(+a)$; arithmetically this is equivalent to $(-a) \times (-1) = a$. Our scheme works very well so far.

Following in the steps of the great German mathematician Gauss, let us see what interpretation we can give to a rotation not of 180° but of only 90°. What is our $(+a)$ now, standing as it does at right angles to our first number-line? It is reasonable to say that the absolute value of this number has not changed since the length of the section it represents is always a radius of the circle. We know that the absolute value of $(-a)$ is the same as that of $(+a)$, namely $|a|$. We should hardly expect a change of value for positions of a in between 0° and 180°. But what of the sign in front of a? Is it reasonable to write plus after turning the arm through 90°? If we do, why should turning through the next 90° change the sign to minus? Provisionally we are going to mark the upward line positive and the downward line negative, though some further investigation is required.

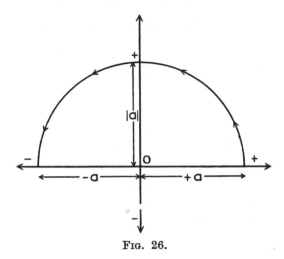

FIG. 26.

We have a means of finding a satisfactory answer to our problem of interpreting a turn of 90°. This is the kind of problem which algebraic equations can solve, so let us try to construct an equation. We begin with the unknown. Let x represent the effect of a 90° turn. Rotating the section $(+ a)$ through 90° gives us therefore $a . x$; rotating it through a further 90° gives us $(a . x)x$. But two rotations of 90° are equal to one rotation of 180° and one rotation of 180° turns $(+ a)$ into $(- a)$. Our equation is therefore

$$(a . x)x = a(- 1).$$

Dividing by a,
$$x^2 = - 1$$
$$x = \pm \sqrt{-1}$$
$$= \pm i.$$

Somewhat to our surprise we find that the turning factor for 90° is the number " i " and that we have found a place for imaginary numbers in our number diagram. The line on which they lie is the line running at right angles through 0 to the line of real numbers. Summing up our findings, we can say that turning $(+ a)$ through 90° gives $(+ ia)$; another turn of 90° gives $i(+ ia)$ or $(+ i^2a)$; but $(i^2) = - 1$, so we now have $- a$ after turning through two quarters of the circle. Turning through the third quarter brings us to $i(- a)$ and a final turn of 90° to $i^2(- a) = (- 1) . (- a) = + a$; we have reached our starting point. These ideas are the basis of what is usually called " the Argand Diagram "—a graphical means of representing real and imaginary numbers. We visualise the numbers, both real and imaginary, as lying along two " axes " at right angles to each other. Notice that our " number-line " has now become a " number-plane " ; for any two lines which meet form a plane or flat surface.

But we are not quite satisfied. We want to use our new knowledge. To this end we will draw the axes of the number-plane afresh, marking them off in numbers.

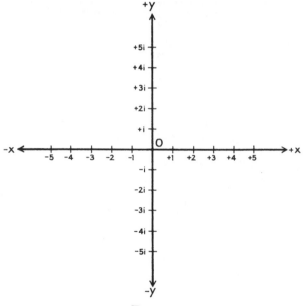

Fig. 27.

We will call the horizontal axis, that is the real number-line, the "x-axis" and its two sections $(+ x)$ and $(- x)$. The vertical axis with its imaginary numbers we will call the "y-axis" ; on the upper section lie $(+ i)$ and on the lower section $(- i)$ numbers. This naming of the axes as x and y is a convention which is adopted for all systems of axes, for whatever purpose they may be drawn. We shall meet them very frequently.

There is one point of interest about the line of imaginary numbers which we should consider before we proceed much further. We have already seen that the line of real numbers is " dense," that is to say that a real number corresponds to any point on it we may choose. There are no gaps however small in the line of real numbers. Are there any gaps in the line of imaginary numbers ? We ought to find the answer to this question now because we are shortly going to look for an interpretation, in terms of number, of points which lie in the plane of the axes but not on the axes themselves. We approach the problem in the following way :—

Our number i is essentially only a kind of " command," namely to multiply by $\sqrt{-1}$. With every number a there is associated its absolute value $|\ a\ |$, whatever kind of number a may be, whether integral, fractional or irrational. This number $|\ a\ |$ can always be found in the positive section of the line of real numbers. In our development of the theory of numbers this part of the number-line was the original idea from which all the rest followed. First we placed the natural numbers on this line ; then we found room for fractions and for irrational numbers.

Now, however, the question of signs becomes very important. In algebra it is only necessary to give $|\ a\ |$ the appropriate sign, $+$ or $-$, to get $(+\ a)$ or $(-\ a)$. Our new axes enable us to interpret the commands $(+\ ia)$ and $(-\ ia)$. The $(+\ i)$ command says " go *upwards* from the zero point as far as $|\ a\ |$." The $(-\ i)$ command says, " go *downwards* from the zero point as far as $|\ a\ |$." The first step in our problem is achieved ; the rule that $|\ a\ |$ is the measure of the distance of the point a from the zero point holds for all the axes. There is therefore little doubt about the " density " of numbers on the imaginary axis ; it is exactly similar to that of the real axis. We have only to remember that i can be used as a coefficient attached to any real number, *e.g.* the numbers $5i$, $\tfrac{3}{4}i$, $38i$, $\sqrt{3}i$, etc., can be thought of as $i5$, $i\tfrac{3}{4}$, $i38$, $i\sqrt{3}$, etc.

Now let us ask the question " How can we add and subtract imaginary numbers from real numbers ? How can we represent numbers of the form $a \pm ib$, so-called complex numbers ? " The similarity of the two axes should now be clear. The turning factors $(+\ 1)$ or $(-\ 1)$ applied to the number a means that the number is to be placed on the real axis; the turning factors $(+\ i)$ or $(-\ i)$ place the number on the imaginary axis. The most general kind of complex number can therefore be written in the form

$$(\pm\ 1)\ .\ |\ a\ |\ \pm\ (\pm\ i)\ .\ |\ b\ |.$$

Now the whole difference between real and imaginary numbers depends on whether $|a|$ or $|b|$ is different from zero or not. If a is zero ($= 0$) then only $(\pm i) |b|$ of our complex number is left and this is the imaginary number $(\pm ib)$. If b is zero ($= 0$) then only $(\pm 1) |a|$ is left, that is the real number $(\pm a)$. When both a and b are zero then we are simply left with 0, zero. (From this point of view we can see that all numbers originate from 0 ; perhaps this is why the meeting point of the axes is usually labelled with the capital letter O ; O = the origin.) If on the other hand both $|a|$ and $|b|$ are different from zero, then the formula $(\pm 1) |a| + (\pm i) |b|$ embraces all possible complex numbers. If $|a|$ and $|b|$ can have any value whatever including zero, then our formula represents in the most general way the whole realm of numbers, real, imaginary or complex.

In order to represent complex numbers on our diagram we must ask ourselves what is meant by the command $a + ib$ or $a - ib$, etc. The first number a means that we are to move along the positive side of the real number axis for a distance a. The second number $+ ib$ means that at the same time we are to move upwards on the vertical axis for a distance b. To take a concrete case ; to represent the number $3 + 4i$ we should have to move, from O, 3 to the right and 4 vertically upwards. In Fig. 28 the point P represents the completion of the two movements ; it represents the number $3 + 4i$.

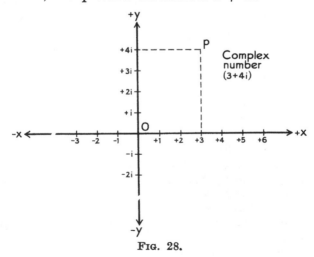

FIG. 28.

We have solved our problem. The complex numbers lie anywhere in the number-plane except on the axes themselves. To illustrate the representation of complex numbers we will give examples in Fig. 29, showing them in all four quadrants.

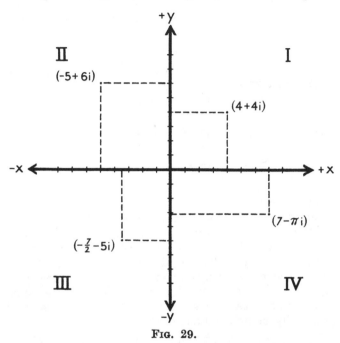

FIG. 29.

Notice the number represented in the fourth quadrant. It is $7 - i\pi$ where π is the irrational number $3 \cdot 1415926 \ldots$

However fruitful and interesting the pursuit of complex numbers would be, leading as they do to the highest realms of pure mathematics, the Theory of Functions and the Theory of the Complex Variable, we must not linger. We will confine ourselves to noting that, for us, these complex numbers are simply algebraic expressions with more than one term and that we can use them in calculations involving the four basic operations of arithmetic. In the last analysis the i is merely another " apple," so long as we note that i is just $\sqrt{-1}$.

However, before we leave the subject of complex numbers we will give a foretaste of the curious properties of this number i. If i is multiplied by itself time and time again we shall find a cyclical pattern which corresponds to its property as a turning factor. For :—

$$i^2 = (\sqrt{-1})^2 = -1$$
$$i^3 = i \cdot i^2 \qquad = i \times (-1) = -i$$
$$i^4 = i \cdot i^3 \qquad = i \times (-i) = -i^2 = +1$$
$$i^5 = i \cdot i^4 \qquad = (+1) \times i = +i$$
$$i^6 = i \cdot i^5 \qquad = i \times (+i) = i^2 \qquad = -1,$$

etc.,

or, in general terms,

$$i^{4n} \qquad = +1$$
$$i^{4n+1} = +i$$
$$i^{4n+2} = -1$$
$$i^{4n+3} = -i$$
$$i^{4n+4} = +1,$$

etc.,

where in this series n is any natural number, 1, 2, 3 . . . etc.

Finding the square and higher roots of complex numbers is more difficult, too difficult for us to investigate in a serious manner, although here is one interesting result that can be verified by squaring both sides of the identity. (An "identity" in algebra shows two ways of expressing the same quantity, e.g., $a + b = b + a$ is an identity.) The square root of a complex number can be split into separate parts, real and imaginary, by using the identity

$$\sqrt{a \pm ib} = \sqrt{\frac{\sqrt{a^2 + b^2} + a}{2}} + i\sqrt{\frac{\sqrt{a^2 + b^2} - a}{2}}.$$

Using this formula we can find a value for the square root of i, for all that we need to do is to put $a = 0$ and $b = 1$. We then have

$$\sqrt{i} = \frac{1}{\sqrt{2}} + i \cdot \frac{1}{\sqrt{2}}$$

and

$$\sqrt{-i} = \frac{1}{\sqrt{2}} - i \cdot \frac{1}{\sqrt{2}}.$$

To find that the square root of an imaginary number is no longer a pure imaginary number need not surprise us now that we realise that it merely means that $\sqrt{i}$ does not lie on the imaginary axis but in the number-plane. (Where ?)

The formula for finding the roots of complex numbers which we have just used is not a very convenient one. There is a much more general formula which is

$$\sqrt[n]{i} = \cos\left(\frac{90}{n}\right)^{\circ} + i \,.\, \sin\left(\frac{90}{n}\right)^{\circ}$$

where n can be any number as big as we like. The extraordinary properties of the imaginary numbers could not be better illustrated. An nth root of i gives a complex number in which a and b are trigonometrical functions!

We are now going to use our hard-won knowledge of complex numbers to bring together several apparently diverse branches of mathematics. Our next study is of " co-ordinates " and " analytical geometry," a subject whose history begins with the Greek geometer Apollonius of Perga and whose main development was due to pressure in the fourteenth century, to the German astronomer Kepler, and to the philosopher Descartes in the seventeenth century.

CHAPTER XXII

CO-ORDINATES

In accordance with our usual practice we are going to construct a machine to help us to grasp the main ideas of our new subject. We are going to deal with points and lines—*mathematical* points and lines, that is to say with points which have position only and no size, with lines which have length but no thickness. These points and lines are such as we can think of but never hope to see. But now to our machine. First we take a drawing board and pin a sheet of drawing paper to it. Then we place on it a T-square so that we can draw a straight line across the bottom of the paper. Now we need a collaborator. His task is to push the T-square up the board at any speed he likes while we draw a line along the edge of the T-square, from left to right, at a steady speed.

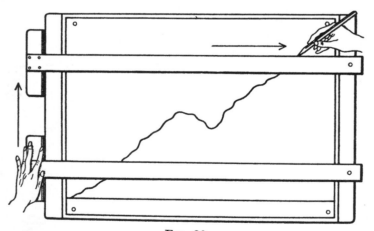

Fig. 30.

The result of our experiment is a rather irregular line on the paper. It goes upwards and from left to right but in a very unsteady manner. In fact at one place where our collaborator allowed the T-square to slip back, the line falls a little before making its final climb. It is clear that if we repeated the experiment we should get a different line and by arrangement with our helper we could perhaps agree to produce some particular kind of line, perhaps even a straight line—but more of this in a moment.

Let us analyse what happened during our experiment. We know that the pencil point was drawn at a steady speed from left to right and at the same time it was rather erratically pushed up the board. At any instant during the experiment the pencil was recording these two distinct movements which were at right angles to each other. If a train runs into a rainstorm in which the rain is falling vertically, the splashes on the window will not be seen as vertical but as slanting traces. The faster the train goes, the more the traces slant. These traces are recording two movements at right angles to each other and taking place at the same time.

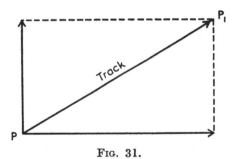

FIG. 31.

The diagram of Fig. 31 is an example of what is known as the " Parallelogram of Velocities." It shows the result of the two movements which the point P makes. It moves both horizontally and vertically as required by the two separate motions, one to the right and one upwards. The straight line, which represents the track of the moving point P is of course an ideal. In our experiment we could never hope to achieve the two uniform speeds which are needed to produce such a result. Any line drawn in a practical experiment would be more like that of Fig. 32.

To analyse such irregular lines mathematically we shall have to look at them from a different aspect. They will have to be considered as a series of points on a line ; but of course they are mathematical points which have position but no size.

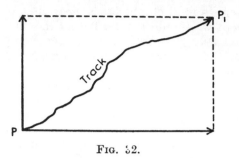

FIG. 32.

In Fig. 33, P represents such a point on the track of the moving pencil.

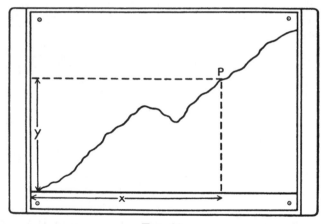

FIG. 33.

We can say that, as a result of the two movements, the pencil on reaching P has moved a distance x to the right and a distance y upwards. For every point on the line there will be a distance x and a corresponding—or co-ordinated—distance y. These two numbers associated with any point of the line are called the " co-ordinates " of the point.

This does not take us very far. The tracks we have drawn contain an infinity of points ; we cannot possibly measure the x and y of every point. It should be clear that there is an infinity of points on the track for, as we kept the pencil on the paper all the time, there is one point on the track corresponding to each point on the base line. And the base line we

may take to be the real number-line. So we can select points
on the base line which correspond to fractions or even irra-
tionals and above any such point we shall find a point on the
track.

It is clear now that what we need is a formula which will
tell us the corresponding values of x and y. We know however
that such a formula is merely an indeterminate equation with
two unknowns, though we can at least choose the values of x.
If we substitute these values of x in the formula, then the
corresponding values of y calculated from it should follow if
the formula is correct.

There is another way of looking at the problem. What we
need is a function, that is a relation between x and y which
holds good for every point of the track. But the prospect of
finding this is rather bleak. If the track were a straight line,
part of a circle or even an ellipse, we might reasonably hope
to find the track function. But if the irregular zig-zag line
we have drawn could be represented by a function it would
be a most unpleasantly complicated one. There is a last
possibility. We might think of breaking up the irregular
track into very small but regular pieces and then find the track
functions corresponding to each piece. This again is rather
hopeless, but if it is generally true that for any track there is
a corresponding function, we might reasonably expect that for
any function there is a corresponding track. We are going to
give some consideration to this kind of analysis—" analytical
geometry " as it is called—because it will provide us with
another tool for our mathematical kit-bag.

First of all let us remind ourselves of what a function is. It
is written generally as $y = f(x)$ where $f(x)$ stands for some
arrangement of x's and constants. For example,

$$y = (2x + 5) \sqrt{5x - 3}$$

is a function of x, though a rather complicated one. Similarly
$y = 5x + 13 \sin x$ is another function of x. It would be as
well to notice here that the word " function " has two meanings. Firstly, it is sometimes understood to stand for the
whole equation and secondly, sometimes for the value of y
itself and the values it can take as we change the values of x.
We could write

$$(\text{Function of } x) = f(x) = 5x + 13 \sin x$$

or more clearly

$$y = f(x)$$
$$y = 5x + 13 \sin x$$
$$\overline{f(x) = 5x + 13 \sin x.}$$

Of course it could go this way :—

$$f(x) = y$$
$$f(x) = 5x + 13 \sin x$$
$$\overline{y = 5x + 13 \sin x.}$$

To make the idea of a " function " quite clear and to show that the two meanings of the word are reconcilable, let us consider the equation

$$15x^2 + 9x + 3y = 12x - 27.$$

In this equation both x and y appear but x is in two powers and y has a coefficient, namely 3. If we try to use this function as it stands we meet difficulties because it is not a straightforward function ; before we do anything else we must straighten it out. A function like this is said to be " implicit " because we cannot directly find the value of y for any value of x ; the value of y is only implied. To make the function " explicit " we have to re-arrange the terms by solving it for y, i.e., isolating y as we isolated x when solving simple equations. We treat the expression as if y were the unknown and the x's were constants. The equation becomes a function only when we are allowed to substitute for x any numbers we like. To make our function explicit we solve the equation for y in this way :—

$$15x^2 + 9x + 3y = 12x - 27$$
$$3y = 12x - 27 - 9x - 15x^2$$
$$3y = 3x - 15x^2 - 27$$
$$y = x - 5x^2 - 9$$

or, arranged in powers of x, $y = -5x^2 + x - 9.$

Now we can say that y is a function of x, or $y = f(x)$, if we remember that x can have any value we care to choose. For every new value of x we shall, in general, obtain a new value of y. We will demonstrate this practically by giving a table of results.

x	y	x	y	x	y
1	-13	$\frac{1}{2}$	$-\dfrac{39}{4}$	0	-9
2	-27	$\frac{2}{3}$	$-\dfrac{95}{9}$	$\sqrt{2}$	$-17\cdot586$ ..
3	-51	$\frac{1}{8}$	$-\dfrac{573}{64}$	π	$-55\cdot2064$ ..
4	-85	$\frac{4}{7}$	$-\dfrac{493}{49}$	e	$-43\cdot227$ ..

In the first column of the table we substituted positive
whole numbers for x ; in the second we substituted fractions ;
and in the third we used 0 and some irrational numbers. In
every case we were able to find a definite value for y. For
every x there is a y, so that for every substitution we get a
" number-pair."

In the expression $y = f(x)$ such as we have just made explicit,
what kind of a number is y ? Though x is any number, the
value of y depends on the arrangement of the x's and 'other
numbers in the "$f(x)$." The value of y is therefore tied to
the value of x, it is limited by or " dependent " on x. If we
give x a series of values which change by small steps we find
that y follows a certain sequence and this sequence is determined by the formula. As x " moves " so y also " moves,"
though not necessarily in the same way as x. If we think of
this movement and try to picture it, we can see it as a " track "
on the drawing board. It is a curve—the graph of the function.

As we have already said, we can reverse the process. We
might just as well say that a sequence of number-pairs is a
function. From this it follows that a function can have three
forms :—

(1) An implicit or explicit equation with two unknowns,
e.g., $y = -5x^2 + x - 9$. (We are confining ourselves to two
variables.)

(2) A curve for which we have to find the formula or
equation.

(3) A table of number-pairs such as might be observed in a
scientific experiment.

In the second case, as we have already said, the function has
to be sought. In the third case it has first to be established

whether there is a functional relation, that is, whether the numbers do follow some law.

How can we change a given function into a curve ? This is the next question to be answered. The question of how a curve can be converted back into a function will be dealt with in the last chapter, together with the problem of getting a curve from a sequence of number-pairs (the problem of interpolation).

We ought now to put our ideas about co-ordinates into some kind of formal order. It will not be difficult because we have in fact been using them in diagrams representing imaginary numbers. We have been using the Cartesian (*i.e.*, of Descartes) or orthogonal (*i.e.*, right-angled) system of co-ordinates which is perhaps the simplest. It is shown in Fig. 34. t h

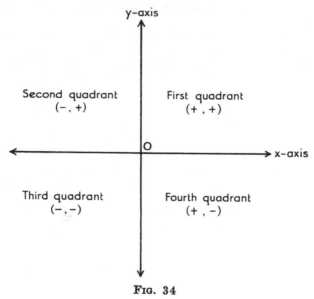

FIG. 34

The point O is called the " origin." The *x*'s are called " abscissæ " and the *y*'s " ordinates." The *x* and the *y* of any point are together called the " co-ordinates " of the point. The " quadrants " are the quarters into which the planes of the axes are divided ; they are numbered in a clockwise direction from 1 to 4. To explain the signs in the quadrants we think of both the axes as being lines of real numbers. To the left of the vertical line the x's are negative ; below the horizontal line the y's are negative.

With this scheme we can find a point corresponding to any pair of real numbers. It should be emphasised that this system cannot be used simultaneously for complex numbers. If we wish to represent both real number-pairs and complex number-pairs then we must have two different planes. But this brings us to the subject of " conformal representation " which is far beyond the scope of this book.

CHAPTER XXIII

ANALYTICAL GEOMETRY

IT should not be forgotten that our aim in these present
chapters is to cover the groundwork which is necessary for an
understanding of the infinitesimal calculus. There are many
subjects which we have touched upon lightly which, though
interesting in themselves, do little to further this aim. We
are therefore going to examine Analytical or Co-ordinate
Geometry in a very sketchy manner, although a thorough
grasp of this subject is essential to a full understanding of
higher mathematics. Now let us return to our task.

Though it may appear to be irrelevant, the first problem is
to consider the diagram of Fig. 35. Here, vertical lines of

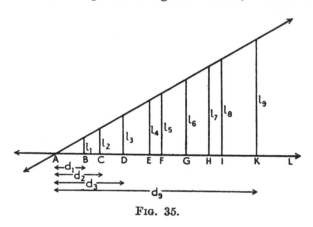

FIG. 35.

different lengths are drawn upwards from the base line. Under
what conditions do the end points of these lines themselves lie
on a straight line ? As is often done in geometry, we will
assume that they do lie on a straight line and then we will
look for the necessary conditions.

In the diagram the points B, C, D . . . K are the points on
the base line L from which the perpendicular lines $l_1, l_2 \ldots l_9$
are drawn. The ends of these perpendiculars lie on the sloping
line L_1. The distances between the ends of the perpendiculars

are $d_1, d_2 \ldots d_9$. As the right-angled triangles which have one vertex at A are all the same shape we know, by a theorem in elementary geometry, that their corresponding sides are proportional. In other words, we know that

$$\frac{l_1}{d_1} = \frac{l_2}{d_2} = \frac{l_3}{d_3} = \ldots = \frac{l_9}{d_9}.$$

This is the mathematical way of saying that all the ends of the perpendiculars lie on a straight line.

Suppose that the ratio $\frac{l_1}{d_1} = k$, a number which may be integral or fractional. Then we can say that for any one of the right-angled triangles

$$\frac{\text{Perpendicular}}{\text{Base}} = k.$$

or $\qquad$ Perpendicular $= k \times$ Base.

If we now refer to a co-ordinate system and make A the origin we can write y for the perpendicular and x for the base. We then have $\qquad y = kx,$

which is the general mathematical way of stating that all the ends of the ordinates corresponding to any values of x lie on a straight line.

Any straight line is known and fixed if we know two points on it. Let us test our result by taking a numerical example in which $k = 2$. We can now choose two values of x, let us say $+3$ and -2. In Fig. 36 the co-ordinate axes are shown and we have to find exactly where in the figure our line should be drawn.

Our condition for a straight line is $y = kx$. When $k = 2$, the condition becomes $y = 2x$. If $x = 3$, $y = 6$ and if $x = -2$, $y = -4$; points corresponding to these number-pairs are shown in the diagram. It will be seen that the line joining them passes through O, the origin. If the reader copies this diagram on graph paper, the value of y corresponding to any chosen value of x can be plotted and the points will always be found to lie on the line. We have in fact found the "analytical equation" of a straight line as the function $y = kx$. In this equation both the variables are of the first power or degree and that is why this type of equation (or function) has been named "linear." Before we continue, we ought to make the difference between the words " function " and " equation " clear. Every function is an equation because it is written formally as $y = f(x)$. But every equation is not necessarily a function. Obviously $5x^2 + 3x + 9 = 27$ is an equation but it is by no means a function. There is only one

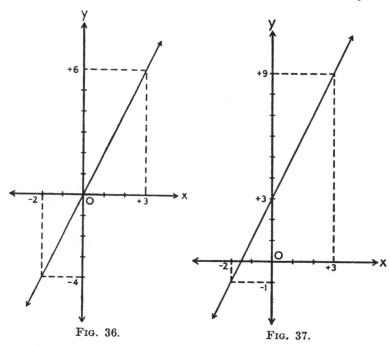

FIG. 36.   FIG. 37.

" unknown " in it, x, and moreover we are not free to choose its values.

We have not yet finished with the equation of the straight line, however. We defined a linear function as being one which produces a straight line and in which the variables are of the first degree. Surely the equation

$$y = 2x + 3$$

should then produce a straight line. It contains a constant term but is otherwise not unlike our example $y = kx$. Let us test it and show the results in Fig. 37.

For $x = 3$, $y = 2 \times 3 + 3 = 9$ and for $x = -2$, $y = 2 \times (-2) + 3 = -1$. The straight line joining these points does not pass through the origin, O, but cuts the y-axis

at ($+$ 3). We could have discovered this by calculation, for
when $x = 0$ the equation shows us that $y = (+\ 3)$. In other
words, we have asked ourselves the question " where does the
line cut the y-axis ? " Similarly, to find where the line cuts
the x-axis we put $y = 0$ and our equation then gives $0 = 2x + 3$,
so that $x = -\frac{3}{2}$. A glance at the diagram shows that this is
indeed the case. Our new technique is evidently a happy
combination of arithmetic and geometry. Let us test it further.
We have already stated that the general form of the straight-
line equation is
$$y = kx$$
where $k = \dfrac{\text{Perpendicular}}{\text{Base}}$, i.e., the perpendicular and base of
any right-angled triangle in Fig. 35. Since these are the
sides of a right-angled triangle there must be a trigonometrical
relation between them and the angle α of the triangle (Fig. 38).

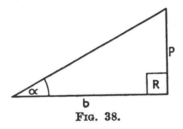

FIG. 38.

If we refer back to our definition we can see that
$\dfrac{\text{Perpendicular}}{\text{Base}}$ is the tangent of the angle α. Therefore
$k = \tan \alpha$. So, in the equation
$$y = kx,$$
the coefficient of x is always the tangent of the angle α.
Therefore in the explicit linear function of the form $y = kx + c$,
where c is a constant, the k is always the value of the tangent
of the angle α, that is to say, of the angle between the line and
the x-axis. The constant c never affects the angle as can be
verified by drawing. It merely pushes the line up or down
the y-axis without altering the angle. If we know the tangent
of the angle we know also the angle itself. So we do also
know the angle which our line makes with the x-axis, that is
the slope or inclination of the line. We use these important
facts later on.

Now that our appetite for analytical geometry has been whetted, we may well consider a curvilinear figure, for example a circle. In short, we want to find a formula which gives for

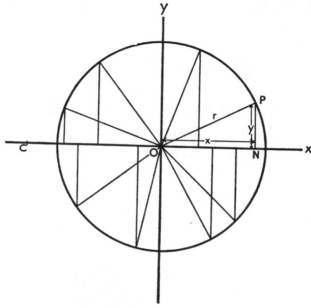

FIG. 39.

any value of x an ordinate, or value, of y, the end of which lies on a circle. Let us begin by drawing the co-ordinate axis. With the origin of co-ordinates as the centre we will now draw a circle of any radius and call the radius r. Next we mark any point P on the circumference of the circle. We join P to O, the origin, and draw the perpendicular PN to the x-axis. Now, using our co-ordinate geometry symbols, the length ON can be labelled x, the length PN will be the corresponding y. The length OP is obviously r, the radius of the circle. Since the triangle OPN is right angled, we can apply the theorem of Pythagoras from which we write

$$r^2 = x^2 + y^2$$
or
$$y^2 = r^2 - x^2$$
or
$$y = \pm \sqrt{r^2 - x^2}.$$

We should notice that the value of x cannot be greater than

r because, if it is, the value of y is imaginary. A glance at the diagram shows that a perpendicular drawn from a point on the x-axis outside the circle does not meet the circle at all. We can note also that for any particular value of x there are two possible values of y, one positive and one negative, a fact that is again explained by reference to the diagram. The surprising thing about our equation is that its algebraic characteristics, e.g., the two signs, the limits on the value of x, can be so simply interpreted in terms of co-ordinate geometry.

Let us take another bite at the analytical bun. We are going to try to solve a quadratic equation by using in rather a neat way the equation for the circle. We saw that the equation for the circle can be written

$$y^2 = r^2 - x^2.$$

If $y = 0$, then $\qquad 0 = r^2 - x^2$

or $\qquad\qquad\quad x^2 = r^2$

or $\qquad\qquad\quad x = \pm\, r.$

Analytically we have found here that when the ordinate (the y) is zero, then x is equal to $+\,r$ or $-\,r$; that is, this condition is met at the points P and Q of Fig. 40, i.e., the

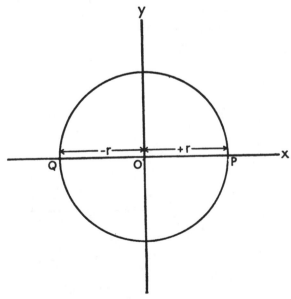

Fig. 40.

points at which the circle cuts the x-axis. The two roots of the quadratic equation $x^2 = r^2$ correspond to the two points at which the circle cuts the x-axis.

We have here touched upon a very important matter. By starting with a function and then putting $y = 0$ we have turned the function into an equation. The reverse process is also used. Suppose we have to solve the *equation* $x^2 - 2x - 15 = 0$. This could be done by drawing the graph of the *function* $y = x^2 - 2x - 15$ and then finding from the graph the abscissæ of the points at which the curve cuts the x-axis, because at these particular points $y = 0$ or, alternatively, $x^2 - 2x - 15 = 0$. If the curve is carefully drawn on graph paper it will be found that it cuts the x-axis at the points where $x = +5$ and $x = -3$. It is hardly worthwhile using this graphical method for solving equations which are as simple as the one we have just used. If the equation can be solved by algebraic methods then an exact result is possible ; the graphical solution is only approximate though, with careful drawing, it may provide solutions accurate enough for many practical purposes.

To summarise ; when using the graphical method we must convert the equation into a function of x, give x a range of values differing by equal steps and find the corresponding y's, plot points on the curve and decrease the steps in x near the points where the curve approaches the x-axis. We shall then get a curve like that of Fig. 41. When $x = 1\frac{1}{2}$ the curve is below the x-axis ;

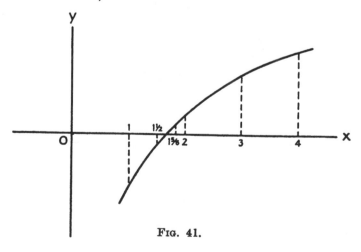

Fig. 41.

when $x = 1\frac{5}{8}$ the curve is above the x-axis, so that the curve crosses the axis at some point in this interval, *i.e.*, a solution of the equation obtained by putting $y = 0$ lies in this interval. By taking values in between we may hit upon the exact position of the cutting point. But more usually we try to narrow down the interval in which the cutting point occurs by taking small steps on either side of the cut. This method of approximation is sometimes called the " regula falsi "—the rule of the false—because we determine the correct value of x by finding two wrong values, one just too big and the other just too small.

We must not linger however because there seems more and more to learn, the further we get with our subject. We can pause only to mention that the number of cuts a curve can make on the x-axis depends upon the highest power of x in the function. For example, a linear function makes no more than one cut, a quadratic function no more than two, a cubic no more than three, and so on. Quite generally we can say that a function of the nth degree makes n cuts, but we may not always find them all upon the graph because some " cuts " may occur at imaginary or complex values of x which cannot be represented on a co-ordinate system of real number-pairs.

For practical purposes we shall include here the algebraic method of solving the general quadratic equation. This equation can be written in the general form

$$x^2 + bx + c = 0.$$

We already know that $(a + b)^2 = a^2 + 2ab + b^2$, and we use this knowledge, this identity, in solving the equation. First we " transfer " the constant to the other side of the equation :—

$$x^2 + bx = -c.$$

Now we use a dodge. We make the left-hand side a perfect square by adding to it the term $\dfrac{b^2}{4}$. This makes the left-hand side $x^2 + bx + \dfrac{b^2}{4}$. By using the identity mentioned before, we can now write the left-hand side as $\left(x + \dfrac{b}{2}\right)^2$. Whatever is added to one side of an equation must also be added to the other in order to keep the " balance." So we have to add $\dfrac{b^2}{4}$

to the right-hand side as well. Therefore

$$x^2 + bx + \frac{b^2}{4} = -c + \frac{b^2}{4}$$

or

$$\left(x + \frac{b}{2}\right)^2 = \frac{b^2}{4} - c.$$

We now take the square root of both sides :—

$$\sqrt{\left(x + \frac{b}{2}\right)^2} = \pm\sqrt{\frac{b^2}{4} - c}$$

or

$$x + \frac{b}{2} = \pm\sqrt{\frac{b^2}{4} - c}$$

and finally

$$x = -\frac{b}{2} \pm\sqrt{\frac{b^2}{4} - c}.$$

This useful formula is valid only when the coefficient of x^2 is 1. If we should have to solve an equation in which x^2 has some other coefficient we must first modify the equation by dividing it by this coefficient. For example, if the equation is $2x^2 + 5x - 3 = 0$, we first divide by the coefficient of x^2 which in this case is 2. Then

$$x^2 + \tfrac{5}{2}x - \tfrac{3}{2} = 0.$$

Comparing this with the general equation

$$x^2 + bx + c = 0$$

we see that here $b = \tfrac{5}{2}$ and $c = -\tfrac{3}{2}$. The equation is solved by substituting in our formula :—

$$\begin{aligned}
x &= -\frac{b}{2} \pm\sqrt{\frac{b^2}{4} - c} \\
&= -\frac{5}{4} \pm\sqrt{\frac{25}{16} - \left(-\frac{3}{2}\right)} \\
&= -\frac{5}{4} \pm\sqrt{\frac{25}{16} + \frac{3}{2}} \\
&= -\frac{5}{4} \pm\sqrt{\frac{25 + 24}{16}} \\
&= -\frac{5}{4} \pm\sqrt{\frac{49}{16}} \\
&= -\tfrac{5}{4} \pm \tfrac{7}{4}.
\end{aligned}$$

So that x is either $-\frac{5}{4} + \frac{7}{4} = \frac{2}{4} = \frac{1}{2}$

or $\qquad\qquad -\frac{5}{4} - \frac{7}{4} = -\frac{12}{4} = -3.$

The curve which has the equation $y = 2x^2 + 5x - 3$ would, if plotted on graph paper, be found to cut the x-axis at the points where $x = \frac{1}{2}$ and $x = -3$.

We will complete this chapter by summarising our results. The function $y = f(x)$ can always be represented by a curve in our system of co-ordinates, but in saying this we are including the straight line as the limiting case of a curve. We often do include limiting cases in mathematics in order to make our results as general as possible. For example, where we considered powers of x we found that we could write $x^2 + 3x + 2$ as $x^2 + 3x^1 - 2x^0$ so that the last term, a mere numeral, could become, if we wished it, a power of x. In this chapter too we saw that the straight line is a curve of the first degree, the circle is of the second degree, as are also all those curves, parabolas, ellipses, hyperbolas, which are obtained by slicing a cone. There are curves of still higher degree but they become of less general interest, the higher they go.

There are many fascinating problems in co-ordinate geometry to which we could give our attention, for instance, finding the intersection of two curves, finding the equations of " tangents " to curves, and so on. But we must press on to problems of higher analysis. In order that we may grasp these properly we will first briefly describe their historical development and discuss them in some detail.

CHAPTER XXIV

SQUARING THE CIRCLE

MOST readers will have heard of the problem of " squaring the circle " and, though they may perhaps not understand what the problem is, they will have gathered that, like the problem of perpetual motion, it is now known to be insoluble. What exactly is the problem ? It is to find a way of dividing up a circle into an *exact* number of small squares, or what amounts to the same thing, to find a square which has *exactly* the same area as a particular circle. Only in the 1880's was it definitely established by Lindemann's mathematical proof that the problem is in fact insoluble. It had of course long been suspected and believed that this was the case. We know if only from the Leibniz series that the number π is an infinite decimal of an irrational kind. As the area of a circle is represented by πr^2, also an irrational multiple of the radius r, it is not surprising that it is impossible to draw a square with sides of $\sqrt{\pi} \times r$ exactly. It might be thought that if we made the squares sufficiently small we might be able to represent the area of a circle exactly. But this in fact never happens ; there is always a part of a square left over. Of course if we make the squares infinitesimally small so that they become points then we do get an exact fit. But we are no better off because we then find we need an infinity of such squares.

It was known however to the great Greek mathematician Archimedes that there are figures bounded by curved lines whose areas are not irrational. We can easily devise a simple " proof " of this. Cut out an inch square of thin cardboard and weigh it on a precision balance in the laboratory. Say it weighs $\frac{1}{2}$ gram. Now cut out from the same sheet of cardboard another figure bounded by curved lines ; trim it until it weighs say exactly 3 grams. If we can assume that the cardboard is of uniform thickness then it follows that the irregular curvilinear figure we have just weighed has an area of exactly 6 square inches. Such an experiment can be repeated as often as we like. There is no hint of irrationality however complicated the figure may be.

It was such consideration of the so-called " lunes of Hippo-crates " which made the Greeks believe that the squaring of the circle was possible. To explain this curious problem of the lunes we have to use an extension of the theorem of Pythagoras. The simplest form of this theorem states that the *square* on the hypotenuse of a right-angled triangle is equal to the sum of the squares on the other two sides. The more general statement of the theorem is that the area of any figure drawn on the hypotenuse of a right-angled triangle is equal to the sum of the areas of *similar figures* drawn on the other two sides. A simple proof of this for the particular case when the similar figures are triangles can be given using Fig. 42.

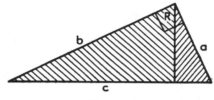

FIG. 42.

This figure shows a right-angled triangle with a perpen-dicular line drawn to the hypotenuse from the opposite vertex. This perpendicular line divides the large triangle into two smaller triangles which, as can be seen by comparing angles, are not only similar to each other but similar also to the original triangle. Now let us think of the figure as a right-angled triangle with similar triangles drawn on its three sides, but drawn inwards from the sides. The triangle on the hypo-tenuse is the original triangle itself ; the similar triangles drawn on the other two sides are the two triangles into which the original was divided by the perpendicular. It is clear that for this particular case of similar triangles we have verified the more general statement of the theorem of Pytha-goras.

In Fig. 43 is shown another version of the extended theorem, this time using semi-circles instead of triangles. The semi-circle on the hypotenuse is drawn inwards, the other two semi-circles (and all semi-circles are similar to each other), are drawn outwards. If we now apply the theorem of Pythagoras we have :—

Area of semi-circle on hypotenuse = Area of triangle + area of segments S_1 and S_2.

Area of the other two semi-circles = Area of the segments S_1 and S_2 + area of the " lunes " L_1 and L_2.

so that, calling the area of the triangle T,

$$T + S_1 + S_2 = S_1 + S_2 + L_1 + L_2$$

from which it follows by subtracting $S_1 + S_2$ from both sides of this equation that $T = L_1 + L_2$

or the area of the triangle is equal to the sum of the areas of the two lunes.

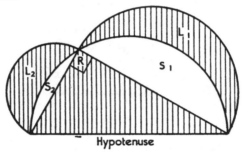

Hypotenuse

FIG. 43.

Since the area of the triangle is easily found from measurements of its sides, it is clear that we have " squared " the two lunes. The lunes however are curvilinear figures bounded by arcs of circles so that we should reasonably expect them to have irrational areas. Yet the diagram of Fig. 43 should convince us that their sum, in any case, has a rational measurable area.

It is now perhaps possible to realise why in classical times the Greeks thought that if only they could find the right approach they would be able to " square the circle." Another problem analogous to that which we have just discussed might be called " cubing the sphere." This problem the Greeks also attempted but they were unable to find any general method of calculating the volume of solids which are bounded by curved surfaces. They could only have recourse to the method which Archimedes used in estimating the quantity of gold in King Hieron's crown—the application of what is now known as the Principle of Archimedes. This method of immersing the solid in a vessel brim full of water and finding how much water is spilt in the process, or the method we have already mentioned of weighing the object, were the only possible ways of finding the volume of any irregular solid. They needed apparatus and some kind of laboratory. What we are hoping to find is a method which will need only pencil and paper to give us the answer.

In the seventeenth century the mathematicians Fermat, Cavalieri, Pascal, Wallis, de Witt, to mention a few, continued to work on the problem of finding areas and volumes ; they came very near a solution by using methods similar to that of Archimedes. Then the light dawned and the problems were first solved by Newton and Leibniz in their invention of the infinitesimal calculus.

Now it is time to bring our history to a close and show step by step how the problem is solved. We shall have to forgo indulging in the philosophical niceties which surround the method we are going to use—indeed these have not yet been satisfactorily dealt with. The expert in pure mathematics will probably hold up his hands in horror at our exposition. We believe however that it is better to have a rough idea than no idea at all. In any case our method was good enough to satisfy the mathematicians of the seventeenth century and any enthusiastic reader who wishes can pursue the subject further in a more comprehensive book of higher mathematics.

We are going to confine ourselves for the present to the problem of finding the area of plane surfaces which are bounded in part by curves, $i.e.$, to the problem of " quadrature." The calculation of volumes will be omitted. Our method requires the use of co-ordinate geometry and so it was developed after Descartes had founded this subject. We will begin by considering the shaded area OBD in the diagram of Fig. 44.

The curve C is not an arc of a circle nor of any other particular curve. It is intended to be merely the arc of any curve of which we know the " equation." Let this equation be $y = f(x)$. By this equation there is a value of y determined for any value of x we care to choose. We are going to leave the equation in this very general form ; we merely assume that $f(x)$ stands for some arrangement of x's and constants. We can substitute any particular algebraic expression for $f(x)$ later—but for the moment the details do not matter.

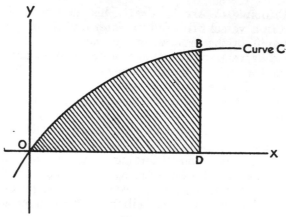

FIG. 44.

In case this very general approach disturbs the reader, let us remind ourselves again of what this " function " is. We are using $y = f(x)$ as a shorthand way of writing

$$y = 3x^2 + 5$$
or $$y = 3x^2 + 4x + 9$$
or $$y = 2\sqrt{x^2 - 1},$$
etc.

All these equations are just special cases of the form $y = f(x)$. In every case if we substitute a number for x we determine a related value for y. This y is merely the arithmetical result of substituting a concrete value for x in $f(x)$. It is then represented as an ordinate standing on the x-axis at the appropriate value of x. The tops of all the possible ordinates constitute the curve or the graph of the function.

We should add one further note of explanation. Broadly speaking, a function expresses a law, a physical relationship between two quantities. When a metal is heated it expands. The expansion of the metal is a " function " of the temperature. We meet thousands of similar examples in everyday life. The speed at which a body falls to the earth is a function of the gravitational force acting on it. The goodness of a wine or of a Gorgonzola cheese is a function of its age.

Here is another famous example. Suppose we could girdle the earth with strips of metal each 1 metre long, that is, put a ring of metal completely round the Equator. We will assume that the earth is smooth and perfectly spherical. Now, having arranged this ring so that it lies flat on the ground at all points, let us insert in the ring another of the 1-metre strips. By how much would one expect the ring to be lifted above the ground ? Most people would estimate that the distance would be minute, perhaps less than a millimetre. But let us see what answer mathematics as opposed to common sense will give. Suppose the radius of the earth is r metres ; then its circumference will be $2\pi r$ and this is the length of the original ring. When we have inserted the additional strip, the length (or circumference) of the new ring is $(2\pi r + 1)$ metres. The radius of a ring can be calculated from its circumference by dividing the latter by 2π. So we have :—

Radius of new ring $= \dfrac{2\pi r + 1}{2\pi}$ metres.

Radius of original ring $= \dfrac{2\pi r}{2\pi}$ metres.

The difference between the two radii is the distance we want, the height to which the ring will be lifted all round the Equator.

Therefore
$$\text{height} = \frac{2\pi r + 1}{2\pi} - \frac{2\pi r}{2\pi}$$
$$= \frac{2\pi r + 1 - 2\pi r}{2\pi}$$
$$= \frac{1}{2\pi} \text{ metres.}$$

To convert this result into something we can more easily appreciate we remember that 1 metre is 100 centimetres, so that

$$\frac{1}{2\pi} \text{ metres} = \frac{100}{2 \times 3 \cdot 14} \text{ centimetres}$$
$$= 16 \text{ centimetres approximately.}$$

It will be seen that our common sense guess was very seriously at fault. Our mathematics tells us something more that is of interest—we did not need to know the radius of the earth. The answer, $\dfrac{1}{2\pi}$ metres, is completely independent of the radius, so that were we to perform the same experiment with the ring round our little finger or even round the orbit of Neptune, the result would be the same, 16 centimetres! In this example the increase of the radius of the ring is *not* a function of the radius as we might have expected.

Now we are ready for the attack. We are going to find the area shown in Fig. 45 which lies between the curve $y = f(x)$,

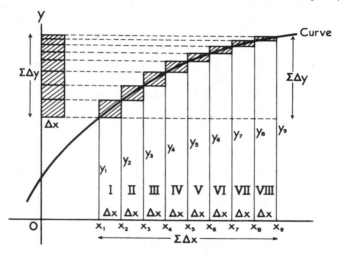

Fig. 45.

the x-axis and the ordinates y_1 and y_9. It is obvious that the area we want lies between the area of all the " outside " rectangles (which includes the shaded parts), and the area of the " inside " rectangles. If the curve were replaced by a straight line all the shaded rectangles would be halved and our task would be simple. As it is, we have to go deeper and look more carefully into the matter. What is the total area of all the inside strips ? Taking them one by one we have :—

$$\text{Area of strip} \quad \text{I} = (x_2 - x_1) \cdot y_1$$
$$\text{,, \quad ,, \quad ,,} \quad \text{II} = (x_3 - x_2) \cdot y_2$$
$$\text{,, \quad ,, \quad ,,} \quad \text{III} = (x_4 - x_3) \cdot y_3$$
$$\text{etc., to}$$
$$\text{,, \quad ,, \quad ,,} \quad \text{VIII} = (x_9 - x_8) \cdot y_8.$$

Now we have arranged the abscissæ $x_1, x_2 \ldots x_9$ at equal distance along the x-axis, so that $(x_2 - x_1) = (x_3 - x_2)$, and so on. As all these differences are equal we can represent them all by the same symbol, Δx (delta x), the "Δ" being understood to stand for "a small part of." We can then write the total area of the inside strips as

$$S_1 = y_1 \cdot \Delta x + y_2 \cdot \Delta x + \ldots + y_8 \cdot \Delta x$$

or more briefly
$$S_1 = \sum_1^8 y_r \Delta x.$$

Since Δx is the same for each term we can apply one of the rules about the use of Σ that we already know and write this formula as

$$S_1 = \Delta x \sum_1^8 y_r.$$

Now, in a similar way, we find the total area of all the outside strips which we will call S_2. From the diagram

$$S_2 = y_2 \cdot \Delta x + y_3 \cdot \Delta x \ldots + y_9 \cdot \Delta x$$
$$= \sum_2^9 \cdot y_r \cdot \Delta x = \Delta x \cdot \sum_2 y_r.$$

We will represent the exact area between the curve and the x-axis by A. Then we know that A is greater than the sum of the areas of the inside strips and less than the sum of the areas of the outside strips. The mathematical way of writing this is

$$\Delta x \sum_1^8 y_r < A < \Delta x \sum_2^9 y_r.$$

The y's of these sums stand for the values which we can obtain from the equation of the curve, $y = f(x)$, if we know the "function of x" for any particular case. Theoretically, we could substitute the nine (or more) values of x in $f(x)$, find the corresponding values of y and then work out the areas S_1 and S_2. We should know that the area we want lies between S_1 and S_2 and we could therefore obtain an approximate value for it. It is clear too that, if we did not object to doing the enormous amount of arithmetical work required, we could make the difference between S_1 and S_2 as small as we like by making Δx smaller, that is by increasing the number of strips. This method is called the Method of Exhaustion—a rather ambiguous title—but it enabled Archimedes to find a surprisingly accurate value for π. However, the calculation required in applying this method to a practical case is immense and at the end of it all we cannot hope to obtain more than an approximate answer.

Mathematicians always avoid heavy computation work when they can, so we may be certain that there must be a way of avoiding the heavy labour which direct summation of all these strips would involve. We have already noted that we get numerical accuracy by decreasing the interval Δx and increasing in a corresponding way the number of strips. Let us take this idea to its logical conclusion. We make Δx so small that it is on the verge of disappearing entirely ; to indicate that we are doing this we change its symbol and write it as " dx." By this move we of course increase the number of strips to " infinity," but at the same time we have found a theoretical way of ensuring that the value of A will be exact. We need now a way of adding together all the y's and multiplying their sum by dx. Notice that we are imagining ordinates standing on every point of the x-axis below the curve ; we have to find their sum! This was the position reached by Cavalieri, who wrote that the problem of quadrature is solved if only we could find " summa omnium y." In 1676 Leibniz made a decisive advance. He wrote : " It will be useful henceforward to write Cavalieri's ' sum of all the y's ' as $\int y dx$."

We have reached the summit of our mathematical mountain. The symbol " $\int y dx$ " is read as " the integral of y with respect to x," a phrase suggested by James Bernoulli. But is this merely word-spinning ? How does it help us ?

We must proceed step by step. Let us first explain the new notation more fully. We have been seeking a formula for the area lying between a curve and the x-axis which is enclosed by two ordinates. Suppose these boundary ordinates stand at the points on the x-axis where $x = a$ and $x = b$. Then the formula we have is the " definite integral " represented by

$$\int_a^b y dx$$

which is read as " the integral of y taken from a to b."

The next step is to consider what the " y " of this formula stands for. It is the y which is equal to $f(x)$ when $y = f(x)$ is the equation of the curve. We can write this :—

$$\int_a^b y dx = \int_a^b f(x) dx.$$

This sign $\int$ is a command—but much as we would like to obey it we do not yet know how to. It is still merely a symbol and we need a " recipe " for it. Let us take a look inside the symbol. It tells us to calculate each of the ordinates under the curve from a to b, to multiply each by dx, a quantity which is almost zero, and then to add up all the results. In other words, we have to carry out this sum :—

1. Area of first strip $= f(a) \cdot dx$
2. „ „ second „ $= f(a + dx) \cdot dx$
3. „ „ third „ $= f(a + 2dx) \cdot dx$
4. „ „ fourth „ $= f(a + 3dx) \cdot dx$

and so on indefinitely, though we know that the area of the last strip but one

$$= f(b - dx) \cdot dx$$
and area of the last strip $= f(b) \cdot dx$.

Here $f(a), f(a + dx)$, etc., are the values of y which correspond to $x = a, x = a + dx, x = a + 2dx$, and so on. The quantity dx is said to be, rather inaccurately, " infinitely small."

Well, Leibniz found the way of performing this enormous sum, as we shall see in Chapter XXXIII.

CHAPTER XXV

CALCULATING THE LENGTHS OF CURVES

WE have used the "infinitesimals" Δx and Δy and the "differentials" dx and dy without much thought for the feelings of the reader. The differentials dx and dy—the "disappearing differences"—are the stumbling block. They are, if we can conceive it, the limiting values of Δx and Δy as these are about to pass into "nothingness." It is at this point that we find it difficult to keep a hold on real, concrete, objective things. Let us comfort ourselves with the thought that when a diamond is found in the blue clay of a South African mine it looks no different from a piece of broken glass. It is not until it has been cut by the skilled craftsmen of Amsterdam that its brilliance, its many facets, become obvious. Much as we may admire the skill of the cutters there would be nothing at all to see if someone had not first unearthed the diamond. Our situation is that we have rediscovered the rough diamond which was first found in the eighteenth century. In the two hundred years since then it has been cut and polished by men like Cesaro, Kowalewski and Peano. We cannot appreciate their achievements until we have had a better look at the diamond as it first appeared when Leibniz and Newton discovered it. Later the reader may, if he wishes, pursue the philosophical aspects of the infinitesimal in more learned text-books of mathematics and eventually come to appreciate the puritanical rigour of its modern developments.

For the present we must return to our diagrams. In Fig. 46 there is a diagram of a curve $y = f(x)$ which rises as x increases. At a point whose abscissæ is x the corresponding ordinate y, or $f(x)$, is drawn. The point on the curve at the end of this ordinate is labelled P. Now we move along the x axis a distance Δx from x and draw the ordinate at that point, that is, where the abscissa is now $x + \Delta x$. The ordinate at this point will of course be higher than the one drawn at x. The increase of y is called Δy so that the ordinate at $x + \Delta x$ is $y + \Delta y$. Let P_1 be the corresponding point on the curve. If we now make Δx smaller and smaller, Δy will also decrease,

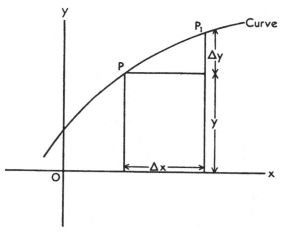

though however small we make our Δx there will always be something left of our Δy as the curve is rising. At the same time P_1 moves down the curve towards P and eventually we can imagine that it bumps into it to form a double point. The line PP_1 is curved but as we make the interval Δx smaller, the line becomes increasingly like a straight line, it is this last property of the diminishing triangle that is the crux of our infinitesimal analysis. We are able to apply to figures with curved lines the results that are already well known for figures with straight lines. We apply these rules both in quadrature and in "rectification," which is the calculation of the lengths of curved lines. To make this application possible we must first have a closer look at an important triangle. Leibniz derived the fundamental ideas of this triangle from the papers of Pascal after the latter's death. Pascal had used the triangle for another purpose but the diagram gave Leibniz the idea he was looking for.

In Fig. 47 A is the point on a curve at which a tangent is drawn. The line AN is the ordinate at A and the line AG is drawn from A at right angles to the tangent. The points B and C are marked on the tangent on opposite sides of A so that AB = AC. The first thing to notice is that the two right-angled triangles BCD and ANG are similar, that is, all their corresponding angles are equal. First, they both have right angles and secondly the acute angle at A is equal to the

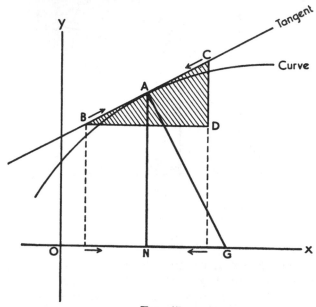

FIG. 47.

acute angle at B because the two lines of the triangle ANG that meet at A are at right angles to the two lines that meet at B. These two pairs of lines therefore form the arms of angles that are equal. If two pairs of corresponding angles are equal then the third pair must be equal because we know that the angles of any triangle always add up to 180°.

Now we first imagine that the points B and C move together towards A, always so that BA = CA. As this takes place the shaded triangle BCD becomes gradually smaller and smaller, though as it does so it retains its shape. Finally B and C will collide with A and the fundamental idea of analysis is within our grasp. At A we have a microscopically small triangle the shape of which we know to be that of the triangle ANG. Though we cannot now see the triangle we know that its angles and the proportion of its sides are the same as those of triangle ANG. We also know that the hypotenuse of the minute triangle is a part of the tangent ; it is also part of the curve at A. It may help the reader to imagine how this is possible if he thinks of a bicycle chain which, though made of rigid straight links, can yet fit snugly round a gear wheel.

Our curve can be thought of as a chain of minute infinitesimal links each of which is straight.

Let us look again at the minute triangle at A, the " characteristic " triangle it is sometimes called. We know that it has the same shape as the triangle ANG. Every point on the curve will have its " characteristic " triangle. It has been drawn, highly magnified, in Fig. 48. All three sides of this triangle are differentials ; the base is dx, the height dy and the hypotenuse we will call ds.

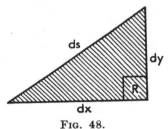

FIG. 48.

We can now think of the curve as being made up of a chain of characteristic triangles, one for each point of the curve. A highly magnified diagram of this chain of triangles is shown in Fig. 49. The length of this apparently broken curve is the sum of all the hypotenuses of the triangles.

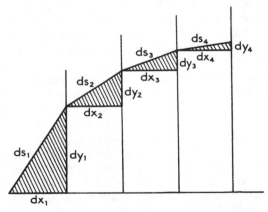

FIG. 49.

We should, however, be misunderstanding the nature of the analysis if we paid too much attention to the characteristic

triangle and forgot the diagram of Fig. 46. We should remind
ourselves that dx_1, dx_2 . . . etc. are steps along the x-axis,
that dy_1, dy_2 . . . etc. are the corresponding increases in the
ordinates and are determined from the equation of the curve
$y = f(x)$ for each value of x. The elements ds_1, ds_2, ds_3 . . .
are the corresponding sections both of the curve and of the
tangents at each point of the curve. A pictorial example may
help. The inverted rocking blotter shown in Fig. 50 is having

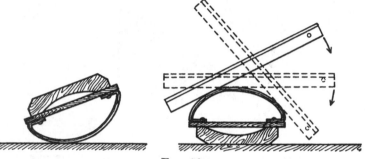

FIG. 50.

a stiff ruler rolled over its blotting surface. The moving ruler
is always a tangent to the surface of the blotter and the point
of its contact with the surface moves along the curve. We
can imagine the curved line as being made up of a chain of
little pieces from the ruler, or conversely, the ruler edge as
being made up of a chain of little pieces from the curved
blotting-paper surface.

The important mathematical idea inherent in this is that by
this breaking down of curves into an infinite train of elemental
straight lines we can apply to the elemental pieces the measurement of straight line figures. This idea is the mathematical
secret for finding the areas of surfaces bounded by curved lines
or indeed of calculating the lengths of curved lines, i.e., of
rectification or " straightening." A curve then is considered
to consist of an infinity of minute links. To find the length of
the curve we find the length of the links and add them up.
The theorem of Pythagoras enables us to find the length of a
link, for

$$(ds)^2 = (dx)^2 + (dy)^2$$

or

$$(ds) = \sqrt{(dx)^2 + (dy)^2}.$$

We find the sum of these as we found the sum of the ordinates in the previous chapter. We find the integral of $\sqrt{(dx)^2+(dy)^2}$ and write it as

$$\int_b^a \sqrt{(dx)^2 + (dy)^2}.$$

As we did before, we put in the limits a and b which are the abscissæ of the ends of the arc we are measuring. All we have to do now is to carry out the integration command, knowing the equation of the curve to be $y = f(x)$. But how ? To find $\int_b^a ds$ does not help us, it merely asks the same question. We should like to be able to obey the command $\int$ but it is as though we are paralysed, rooted to the spot. In the next chapter, however, the way will be shown.

CHAPTER XXVI

DIFFERENTIALS AND INTEGRALS

ALTHOUGH we cannot determine values for the differentials dx and dy, we have shown in Chapter XVIII that if in the equation $y = f(x)$ the exact form of $f(x)$ is known, we can establish the ratio $\frac{dy}{dx}$. Two numerical examples were given. This ratio $\frac{dy}{dx}$ is a very important function : it is called the " derived function " of $f(x)$ or, more simply, its " derivative." Sometimes it is even called its " differential coefficient " ! All the functions we shall refer to in this book will be " differentiable," that is, have derivatives, but it is not generally true that all functions have derivatives ; some are differentiable except for special values of x, some have no derivative at all. The operation for finding the derivative of a function is called " differentiation." The basic rules of this operation, once they are known, are no more difficult to apply than those of multiplication. Our aim in this chapter is to find out whether our knowledge of differentiation can help us to understand how to carry out the integration " command." For this we shall need to change our methods. In recent chapters we have used graphical and geometrical arguments ; for our present purpose we shall find algebraic methods more helpful—just as we did when we found that multiplying by $\sqrt{-1}$ was equivalent to a turn of 90° in the Argand diagram.

In our work on quadrature (Chapter XXIV) we found the following formula for the area that lies " under " a curve :—

$$A = \int y dx = \int f(x) dx$$

where $y = f(x)$ is the equation of the curve. We have also an equation from Chapter XVIII, namely

$$\frac{dy}{dx} = f'(x)$$

where $f'(x)$ is the derivative of $f(x)$. Is there any hidden relation between the formula for area and the derived function ? This is the question which now faces us.

It will help us if, for a moment, we introduce a second function which we call $F(x)$. To accord with our notation the derivative of $F(x)$ will be written as $F'(x)$. We are now going to make, as Leibniz did, a very bold jump indeed. In effect he said :—

If the derivative of $\qquad F(x)$ is $F'(x)$

then the integral of $\qquad F'(x)$ is $F(x)$

or $\qquad\qquad\qquad \int F'(x)dx = F(x)$.

We might restate this in a way that makes it apply more directly to our present problem : the command " $\int f(x)dx$ " says " Find the function of which $f(x)$ is the derivative."

In Chapter XVIII we found that the differential coefficient or derivative of $2x^2$ was $4x$; so that if we write $f(x) = 2x^2$, then $f'(x) = 4x$. We can immediately perform our first integration, namely $\qquad\qquad \int 4x\,dx = 2x^2$.

It is as easy as that ! (But only when we know both $f'(x)$ and $f(x)$.)

Now it is obvious that in the previous paragraphs we have asked the reader to swallow a very indigestible mouthful. If he can stomach it at all, he will perhaps realise two things immediately. Firstly he will see that the acceptance of Leibniz's idea makes the connection between the operations of integration and differentiation surprisingly simple ; integration is the reverse of differentiation, differentiation the reverse of integration. They are reverse processes just as are addition and subtraction, or multiplication and division, or squaring and finding the square root. We shall discover later that we must make some reservations about this reversibility but we need not qualify our statement at present. Secondly, the reader may well say " Yes, I can see it is a very good idea " and be acutely aware that logically we have taken a big jump with no justification whatever. In what sense is the summation command of the symbol $\int$ the " reverse " of the algebraic operations we performed on functions in Chapter XVIII ? Are all functions derivatives of other functions ? There are many searching questions that could be asked.

The only reply to these criticisms that is possible is to say that the theoretical justification of Leibniz's revolutionising idea is difficult and long ; this book is not the place for it. From a practical point of view the step that Leibniz took is justified because, as we shall see in the following chapters, it works. It produces results that are obviously correct and for the present let us be content with that. Elementary " proofs " are sometimes given in text-books, but any critical reader will find that they create more difficulties than they pretend to remove.

Let us repeat the fundamentally important idea which we have introduced in this chapter :

The integral of $f(x)$ is the function of which $f(x)$ is the derivative.

Before we give any further examples of the application of this rule there is one point which we should make at once. The reader may have noticed that when we found the area under the curve or the length of an arc of a curve we used in the final result what we called a " definite " integral, *e.g.*, $\int_a^b f(x)dx$, in which a and b were the values of x which indicated the ends of the curve. In this chapter we have used integrals in which the limits were not stated. Such an integral is called " indefinite." The essential difference between the two types is that the definite integral is merely a particular value of the more general indefinite integral. We can think of an indefinite integral as being a general formula for an area, a length or whatever it may be, just as $\pi r^2 h$ is the general formula for the volume of a cylinder. A definite integral is the result of substituting special values of x in the indefinite integral, just as 314 square inches is the result of substituting $r = 2''$ and $h = 25''$ in the formula giving the volume of the cylinder.

We must turn to the differential calculus for further experience in order to make it possible for us to work out or " evaluate " integrals. We shall perform these operations of differentiation and integration together—a deviation from the usual custom which stresses the inter-relation of the two processes. As a preparation for this work, however, there are two more rivers to cross ; we shall need to know more about " orders of smallness " and about Newton's " Binomial Theorem."

CHAPTER XXVII

THREE KINDS OF SMALLNESS

IN Chapter XVIII we have already had to deal with the small quantities Δx and $(\Delta x)^2$. It may be remembered that we used the argument that if Δx is very small then $(\Delta x)^2$ is smaller still, so small that in comparison with Δx it can be " neglected." We also made $\Delta x \rightarrow 0$ so that Δx was small compared with x. So we have already met two " degrees " of smallness. Physically it is as though we were dealing, for example, with the whole cosmos or universe, the Earth and a speck of dust. Compared with the Earth the value or mass of a speck of dust is " negligible " ; compared with the universe the Earth in turn is negligible. In the infinitesimal calculus we shall frequently be concerned with " orders " of smallness, and it is as well that we spend a few moments now in clarifying our ideas about them.

Suppose we had a 1-inch cube of metal like that illustrated in Fig. 51. We all know that metals expand when heated. Imagine that this metal block is placed on the floor of a room, right up against the walls in a corner. If the metal is now heated we know that it will increase in size and, because it is jammed up against the walls and the floor, its expansion can only take place outwards and upwards. In Fig. 52 we show the expanded cube and also its original dimensions. It can be seen that the expanded cube could be cut up into eight separate parts if we made these clean cuts along its original faces. Of the eight parts, one is the original cube ; three square slabs are S_1, S_2 and S_3 which fit exactly over the original faces ; three more are the long strips L_1, L_2 and L_3 which fit in between adjacent slabs ; and finally there is one piece, P, a small cube which is needed to fill in the top corner. We now allow the cube to cool down almost to its original temperature until the extent of expansion along each of the edges is, let us say, one-millionth of an inch. What would be

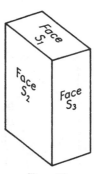

FIG. 51.

the volumes of the separate pieces ? The slabs would have a volume of $1^2 \times 0 \cdot 000001$ cubic inches $= 0 \cdot 000001$ cu. in., the

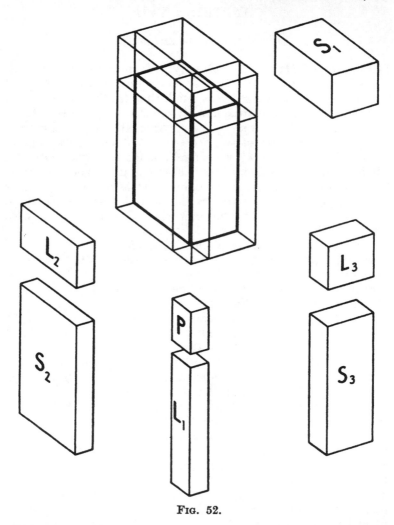

Fig. 52.

strips of $1 \times (0 \cdot 000001)^2 = 0 \cdot 000,000,000,001$ cu. in., and the small cube P of $(0 \cdot 000001)^3 = 0 \cdot 000,000,000,000,000,001$ cu. in. If we allowed the cube to cool a little more so that the expansion is reduced to one ten-millionth of an inch, then the volume of the square slabs would be reduced to $\frac{1}{10}$, of the strips to $\frac{1}{100}$, of the corner cube to $\frac{1}{1000}$ of the values we have just calculated. In other words, as the expansion is reduced to zero the volume of the slabs is also reduced to zero, but the volume of the strips is getting smaller at a much faster rate and the volume of the corner cube is getting smaller at a much faster rate still. If the extent of the expansion is reduced until it almost disappears completely, we can imagine at the last instant that the slabs are so thin as to be almost geometrical planes, the strips would then be almost geometrical lines and the corner cube almost a mere point. Slabs, strips and cube have all become infinitesimally small ; they have almost disappeared yet at the moment just before they pass into zero they still possess all the characteristics of their state in Fig. 52 when the metal block had reached the limit of its expansion. They have now almost reached the limit of their reduction and the slabs are expressed as surfaces only, the strips as lines and the cube as a point. They are barely seen in three dimensions as they are drawn in Fig. 52. They are all infinitesimally small in volume ; but can their smallness be of the same kind ? An infinitesimally thin surface area is almost a plane, an infinitesimally thin line is almost a series of points and a " point " is—just almost a point! These represent different degrees of smallness so that when we speak of Δx, $(\Delta x)^2$ or $(\Delta x)^4$, we mean that if Δx is of the first order of smallness, $(\Delta x)^2$ is of the second order and so on. We can speak of $(\Delta x)^n$ as being of the nth order of smallness and go right up to the limit of zero, that is, just before Δx disappears into nothingness.

We have to recognise ranks or degrees of greatness just as we have now recognised degrees of smallness. It is the latter, however, which are the more important for our present purposes. We shall soon be using them in concrete examples, but before we can do this we must look at the Binomial theorem as we promised. We are going to approach this theorem by way of our already hard-won knowledge of combination ; this will help us to determine the Binomial " coefficient."

CHAPTER XXVIII

THE BINOMIAL THEOREM

WE must begin with a definition. Numbers of the form $(x + a)$ or $(x + b)$ are said to be " binomial numbers," or simply " binomials." If one of these binomial numbers is used as in a multiplication, it is usually called a " binomial factor." The word " binomial " here is used to indicate that the number consists of the sum (or difference) of two other numbers which for some reason cannot be combined more closely. The " x " of the numbers $(x + a)$ and $(x + b)$ is not intended here to represent an unknown number ; it is merely a convenient way of showing that in some respects it can be thought of as a number of a kind different from a and b. In the theory of functions it is often used to represent a variable number while a and b remain constant.

If we multiply together the two binomials $(x + a)$ and $(x + b)$ we have

$$(x + a)(x + b) = x^2 + ax + bx + ab = x^2 + (a + b)x + ab.$$

Similarly,

$$(x + a)(x + b)(x + c) = x^3 + (a + b + c)x^2 + (ab + ac + bc)x + abc,$$

can be written down and left to the reader to verify if he wishes.

We are now going to write down these products in a different way in order to show their structure :—

$$(x + a)(x + b) = x^2 + \left. \begin{matrix} a \\ b \end{matrix} \right\} x + ab \text{ and}$$

$$(x + a)(x + b)(x + c) = x^3 + \left. \begin{matrix} a \\ b \\ c \end{matrix} \right\} x^2 + \left. \begin{matrix} ab \\ ac \\ bc \end{matrix} \right\} x + abc.$$

The practised eye can already see two similarities in these products, firstly that they are arranged in order of the descending powers of x, and secondly the coefficients of the different powers of x are symmetrical arrangements of a, b and c. In the first coefficient after the highest power of x the numbers a, b and c are arranged in ones, in the next coefficient they appear in pairs, then in threes. The kind of arrangement or "complexion" of each of these coefficients is that of a "combination without repetition" of which the elements are the quantities a, b, c.

To make the structure quite clear we will multiply out five different binomials :—

$$(x + a)(x + b)(x + c)(x + d)(x + e) =$$

$$= x^5 + \left.\begin{matrix} a \\ b \\ c \\ d \\ e \end{matrix}\right\} x^4 + \left.\begin{matrix} ab \\ ac \\ ad \\ ae \\ bc \\ bd \\ be \\ cd \\ ce \\ de \end{matrix}\right\} x^3 + \left.\begin{matrix} abc \\ abd \\ abe \\ acd \\ ace \\ ade \\ bcd \\ bce \\ bde \\ cde \end{matrix}\right\} x^2 + \left.\begin{matrix} abcd \\ abce \\ abde \\ acde \\ bcde \end{matrix}\right\} x + abcde$$

The structure should now be quite clear. Each term in this particular product consists of five numbers multiplied together. The first term x^5 is only $x \cdot x \cdot x \cdot x \cdot x$ and ax^4 is $a \cdot x \cdot x \cdot x \cdot x$. Again, $acex^2$ is $a \cdot c \cdot e \cdot x \cdot x$. Another look at the product should convince the reader that the coefficients are combinations of a, b, c, d, e.

We will now summarise our present knowledge and decide how to write down the product of any number of different binomial factors, $(x + a)$, $(x + b)$ . . . etc. The first point is to notice that the x's are set out in a descending sequence—all the powers of x that will arise in the product. If there are "n" factors, then the highest power of x will be x^n; the lowest will always be x^0. So that, without the coefficients, the sequence of powers of x will be

$$x^n + x^{n-1} + x^{n-2} \ldots x^2 + x^1 + x^0.$$

Note that the number of terms is one more than the number of binomials in the product. Next we write down this series again, this time inserting the coefficients :—

$$c_0 \cdot x^n + c_1 \cdot x^{n-1} + c_2 \cdot x^{n-2} + \cdots$$
$$c_{n-2} \cdot x^2 + c_{n-1} \cdot x^1 + c_n \cdot x^0.$$

The right-hand suffix of the coefficients c_1, c_2, etc., shows that the n letters are combined firstly one at a time, then two at a time, and so on. The first term has the coefficient c_0 ;

only x's appear in the term and this coefficient is 1. The terms are real products. The term $c_1 x$ stands for $(ax+bx+cx+ \ldots)$ or $(a + b + c \ldots)x$, similar to the structure we illustrated in the first few paragraphs of this chapter. It is interesting to note that this kind of combination does arise in multiplication.

Next we do another mathematical dodge. Having used the different quantities a, b, c . . . to our advantage we now dispense with all but one of them by putting $b = c = d = \ldots a$, all of them equal to a. Then the binomial product becomes $(x + a)(x + a)(x + a)$ . . . with n factors, which can be written more shortly as $(x + a)^n$. We now have

$$(x + a)^n = c_0 . x^n + c_1 . x^{n-1} + c_2 . x^{n-2} + \ldots$$
$$+ c_{n-2} . x^2 + c_{n-1} . x^1 + c_n . x^0.$$

We have decided that $c_n = 1$. What are c_1, c_2, etc. ? It is now straightforward ; the coefficient c_1 is the sum of the combinations of n different quantities taken one at a time, after which all the different things are put equal to a. So that here $c_1 = (a + b + c + \ldots) = (a + a + a \ldots) = na$ or $_nC_1 . a$. The coefficient c_2 is similarly the sum of n different quantities taken two at a time, and each of which is then put equal to a; so that $c_2 = {}_nC_2 . a^2$. Similarly $c_3 = {}_nC_3 . a^3$, and so on. The required " expansion " can now be written down :—

$$(x + a)^n = {}_nC_0 . x^n + {}_nC_1 . a . x^{n-1} + {}_nC_2 . a^2 . x^{n-2} +$$
$$\ldots + {}_nC_{n-1} . a^{n-1} . x + {}_nC_n . a^n.$$

For small values of n these coefficients can be found without the labours of calculating them by referring to Pascal's Triangle, a table of numbers which can be constructed very simply. The first seven rows are shown below :—

0							1						
I						1		1					
II					1		2		1				
III				1		3		3		1			
IV			1		4		6		4		1		
V		1		5		10		10		5		1	
VI	1		6		15		20		15		6		1
VII	1	7	21		35		35		21		7		1 etc.

We do not propose to justify the use of this table here. We are only going to explain how it is constructed and how it is used. Notice that each number is the sum of the two numbers in the row above it which are nearest to it. Thus in row VI the number $15 = 5 + 10$, which are found above 15 in row V. The whole table is built upon this law so that it can be extended mechanically as far as is worth while. Though this table was discovered and used by Pascal, these " binomial coefficients " were used by Stifel (1544) and were known to Chinese mathematicians as early as A.D. 1300.

A simple example will show how Pascal's triangle is used. If we wish to expand $(x + a)^5$ we know that the highest power of x that will occur is x^5. We therefore write down a descending series of powers of x beginning with x^5 and as we do so we take as their coefficients the numbers in row V of Pascal's triangle, namely 1, 5, 10, 10, 5, 1. The expansion is then :—

$$(x + a)^5 = 1 . x^5 + 5ax^4 + 10a^2x^3 + 10a^3x^2 + 5a^4x + 1 . a^5$$

remembering that x^0 is merely 1.

As an example of the use of the combination form of the coefficients let us expand $(x + a)^7$. Here $n = 7$, and

$$(x + a)^7 = {}_7C_0x^7 + {}_7C_1ax^6 + {}_7C_2a^2x^5 + {}_7C_3a^3x^4 + \\ {}_7C_4a^4x^3 + {}_7C_5a^5x^2 + {}_7C_6a^6x + {}_7C_7a^7.$$

We have already shown in Chapter VII how to find the value of these coefficients. Writing them out in full and then simplifying them :—

$$(x + a)^7 = x^7 + \frac{7}{1} . ax^6 + \frac{7 . 6}{1 . 2} a^2x^5 + \frac{7 . 6 . 5}{1 . 2 . 3} a^3x^4$$

$$+ \frac{7 . 6 . 5 . 4}{1 . 2 . 3 . 4} a^4x^3 + \frac{7 . 6 . 5 . 4 . 3}{1 . 2 . 3 . 4 . 5} a^5x^2$$

$$+ \frac{7 . 6 . 5 . 4 . 3 . 2}{1 . 2 . 3 . 4 . 5 . 6} a^6x + \frac{7 . 6 . 5 . 4 . 3 . 2 . 1}{1 . 2 . 3 . 4 . 5 . 6 . 7} a^7$$

$$= x^7 + 7ax^6 + 21a^2x^5 + 35a^3x^4 + 35a^4x^3 + 21a^5x^2 \\ + 7a^6x + a^7.$$

The conclusion agrees with row VII of Pascal's triangle.

We will now work out a few more results. Here is a numerical example. What is $(103)^5$? At first sight this does not seem to require the use of the Binomial Theorem at all. We could of course work it out by direct multiplication, but we can use the binomial theorem to give us the result without the tedium of long multiplication sums. The trick is to think of

103 as $100 + 3$ and then to put $x = 100$ and $a = 3$ in our formula. Using row V of Pascal's triangle,

$$(103)^5 = (100 + 3)^5 = 100^5 + 5 \cdot 3 \cdot 100^4 + 10 \cdot 3^2 \cdot 100^3$$
$$+ 10 \cdot 3^3 \cdot 100^2 + 5 \cdot 3^4 \cdot 100 + 3^5$$
$$= 10{,}000{,}000{,}000 + 1{,}500{,}000{,}000 + 90{,}000{,}000 + 2{,}700{,}000$$
$$+ 40{,}500 + 243$$
$$= 11{,}592{,}740{,}743,$$

a result which could be verified by long multiplication. This trick is frequently used to find approximate values of squares, cubes or higher powers of numbers that can conveniently be split into a large and a small part, e.g. $1 \cdot 05 = 1 + 0 \cdot 05$, and $20 \cdot 04 = 20 \times 1 \cdot 002 = 20 \,(1 + 0 \cdot 002)$.

So far the binomials we have used have consisted of the sum of two terms. The theorem still applies however if the binomial is a difference, say $(x - a)$. The best way of considering this difference is to regard it as the sum of two terms one of which is negative :—

$$x - a = x + (- a).$$

The expansion of $(x - a)$ to any power is then carried out in the usual way. Here is an example :—

$$(x-2)^4 = (x+(-2)\,)^4$$
$$= x^4 + 4 \cdot (-2) \cdot x^3 + 6 \cdot (-2)^2 \cdot x^2 + 4 \cdot (-2)^3 \cdot x + (-2)^4$$
$$= x^4 - 8x^3 + 24x^2 - 32x + 16.$$

The formulæ we have derived in this chapter have been based on the assumption that the power to which the binomial is to be raised, or the " exponent," as the index " n " is called, is always a whole number. The binomial theorem can be used however when n is a fraction and even when n is a negative whole number or a negative fraction, though we have to apply some restrictions in these cases. For such cases the expansion is used in the form

$$(1 + x)^n = 1 + \frac{n}{1} \cdot x + \frac{n(n-1)}{1 \cdot 2} x^2 + \frac{n(n-1)(n-2)}{1 \cdot 2 \cdot 3} x^3 + \dots$$

the series going on without end. The most important restriction to bear in mind here is that in these cases the value of x, or rather of $\mid x \mid$ to allow for the case of $(1 - x)^n$, must be less than 1. Here is an example :—

$$(1 - x)^{-2} = (1 + (-x))^{-2}$$

$$= 1 + \frac{(-2)}{1} \cdot (-x) + \frac{(-2)(-3)}{1 \cdot 2}(-x)^2$$

$$+ \frac{(-2)(-3)(-4)}{1 \cdot 2 \cdot 3}(-x)^4 + \cdots$$

$$= 1 + 2x + 3x^2 + 4x^3 \cdots$$

Notice that Pascal's triangle is of no use for negative exponents.

The restriction that $|x|$ must be less than one is not as serious as might appear at first sight. There is always a trick up the mathematician's sleeve! To find the value of $\sqrt[3]{124}$, for example, we must put it in the form $(1 + x)^n$. Notice that $125 = 5^3$ so that

$$\sqrt[3]{124} = 5\sqrt[3]{\tfrac{124}{125}} = 5(1 - \tfrac{1}{125})^{\frac{1}{3}}$$
$$= 5(1 - 0 \cdot 008)^{\frac{1}{3}}.$$

We can now apply the binomial theorem to the expression $(1 - 0 \cdot 008)^{\frac{1}{3}}$, taking as many terms of it as are required for our purpose and finally multiplying the total by 5. Since the number $\sqrt[3]{124}$ is irrational we can never hope to find its exact value as a decimal but it will be found that the terms of the binomial expansion in this example decrease very rapidly in size. The first three terms when worked out to six decimal places are :—

$$1 - 0 \cdot 002667 + 0 \cdot 000007 \cdots = 0 \cdot 997340.$$

To complete the calculation $\sqrt[3]{124} = 5 \times 0 \cdot 997346$
$$= 4 \cdot 9867$$

to four decimal places.

There is an important difference between the binomial expansions when the power to which the binomial is raised is a whole number and when this power is fractional or negative. In the former case the binomial series came to an end ; it was a " finite series " ; in the latter case it never comes to an end ; it is an " infinite series." We have already met one infinite series, namely $\frac{\pi}{4} = 1 - \frac{1}{3} + \frac{1}{5} - \frac{1}{7} \cdots$

This series is typical of a kind which is said to " converge," that is, the terms steadily get smaller and smaller and at the same time the total sum approaches more and more closely to a " limit," in this case to $\frac{\pi}{4}$. There is another kind of infinite series in which the *total sum* steadily increases, as it obviously does in the series

$$1 + 3 + 5 + 7 + \ldots$$

This series is said to " diverge " ; there is no limit to which the sum steadily tends. The total just gets bigger and bigger without any limit as successive terms are added.

It is by no means always easy to determine at sight whether an infinite series converges or not. For example in the series

$$\tfrac{1}{1} + \tfrac{1}{2} + \tfrac{1}{3} + \tfrac{1}{4} + \tfrac{1}{5} \ldots$$

though the terms steadily decrease, the total continues to increase indefinitely. On the other hand, the series

$$\frac{1}{1 \cdot 2} + \frac{1}{2 \cdot 3} + \frac{1}{3 \cdot 4} + \frac{1}{4 \cdot 5} + \ldots$$

is convergent. As the reader can see, by splitting each term thus :—

$$(\tfrac{1}{1} - \tfrac{1}{2}) + (\tfrac{1}{2} - \tfrac{1}{3}) + (\tfrac{1}{3} - \tfrac{1}{4}) + (\tfrac{1}{4} - \tfrac{1}{5}) + \ldots$$

the total sum becomes closer and closer to 1 as we continue to add term after term.

Infinite series play a very important part in higher mathematics. Every integral can be expressed in this form and in some cases convergent infinite series provide the means by which their numerical values are determined. It was by using convergent series that Archimedes was able to perform some of his miracles of quadrature, as we shall see in the next chapter.

CHAPTER XXIX

IT is to Archimedes that we must grant the honour of having discovered a way of finding the area of plane surfaces bounded by curved lines. His method was so effective that it was not superseded until the end of the seventeenth century.

We are going to examine the method which Archimedes developed, even though we now have a better method at hand, because it is instructive. We shall choose as an example the finding of the area of a plane enclosed in part by the common parabola. This curve was very thoroughly investigated by Archimedes ; not only did he achieve its quadrature, he succeeded also in the more difficult task of finding its centre of gravity, i.e. the point on which it would balance. For our purposes his geometrical approach to the quadrature problem will have to suffice ; the statics problem is too difficult for us.

Archimedes' method was to use what was later known as the " process of exhaustion." It consists of fitting regular rectilinear figures, i.e. simple figures bounded by straight lines, into the curve. We use triangles, squares, etc., beginning with the largest possible. The spaces between the curve and the rectilinear figures are gradually filled up with further triangles, etc., so that the outer boundaries of the rectilinear figures more and more closely approximate to the curve. All this however requires that we should be able to calculate the areas of all the figures we insert. Since the number of figures is infinite, this is only possible if we can find a law which will enable us to express the sum of the areas as an infinite series. Then if we can calculate the sum of this series we shall have determined the area of the figures which were needed to " exhaust " the area enclosed by the curve. We show the beginning of the exhaustion process in Fig. 53, in which the curve is a parabola, a curve which is symmetrical about the thick line, i.e. the " axis," drawn through its middle.

The parabola is cut off by a chord at right angles to its axis to give what is called a " segment." Since the parabola is symmetrical about its axis it is necessary for us only to find the area of one of the two halves. The first triangle drawn is the large right-angled triangle. The hypotenuse of this triangle joins the vertex of the parabola to the point where the chord cuts the curve ; of its other two sides one lies along the axis and the other along the chord. The next triangle drawn is the one which is shaded in the diagram. This triangle

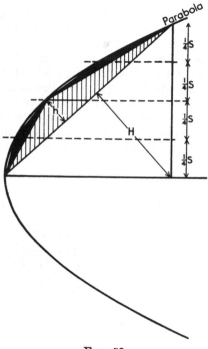

FIG. 53.

has its vertex on the curve and its base is the hypotenuse of the first triangle. There are now two gaps to be filled and we begin by drawing the two triangles which are black in the diagram. So the process continues, always placing the vertices of the triangles on the curve and the bases on the sides of the triangles already drawn. The outline of the straight line figures gradually approaches the curve and we imagine the triangles drawn until at last the fit is exact. But how do we find the sum of their areas ?

The solution of this problem is an excellent example of the clarity and precision of Greek mathematical genius. Archi-

medes knew that infinite series like $1 + \frac{1}{2} + \frac{1}{4} + \frac{1}{8} \cdots$
could be summed ; in this case the total sum approaches the
value 2 as more and more terms are added. In the drawing
of the triangles that were to be used to fill in the curve, Archimedes sought a method of drawing them that would provide
a series of this kind.

Before we can attempt this for the parabola, or indeed for
any other curve, we must know some geometrical properties
of the particular curve. The fundamental property of the
parabola is most simply stated for us in terms of analytical
geometry. If the vertex of the parabola is made the origin
of the co-ordinate system and the axis of the parabola is made
the x-axis of this system, then the parabola has, in its
simplest form, the equation $y^2 = x$ or $y = \pm \sqrt{x}$. This means
that the ordinate of any point on the curve is proportional
to the square root of the abscissa of the point. This basic
property of the parabola was of course known to Archimedes.
One other piece of information which ought to be given here
is that any line parallel to the axis of the parabola is called a
" diameter " ; any other line which cuts the parabola twice
is called a " chord."

We must now refer to Fig. 54. PN is the axis of the parabola
and P′N is the half of a chord at right angles to the axis. The
area of the first triangle we have to consider, the right-angled
triangle PNP′, is $\frac{1}{2} \times$ PN $\times$ P′N $= \frac{1}{2}b \cdot h$. That is straightforward enough, but now we have to decide exactly how we
intend to draw the next triangle. The device Archimedes
used, and one which enables us to solve the problem, is as
follows :—

The line P′N is bisected ; let us call its middle point X_1.
Then NX_1 and $X_1P′$ are each bisected ; we will call these
points of bisection X_2 and $X′_2$. Then again the four quarters
NX_2, X_2X_1, $X_1X′_2$, $X_2′P′$, are each bisected ; these further
points we will call X_3, $X_3′$, $X_3″$, $X_3‴$. This process is continued
indefinitely though it is neither necessary nor desirable to draw
any more of them, as we shall see. The next construction is
to draw diameters through all the X's so that they cut the
curve. To each point X there will then correspond a point P
on the curve. These points P are now made the vertices of
the successive triangles. The diagram shows how this has
been done—following the method which Archimedes suggested.

The next task is to calculate the areas of these triangles.

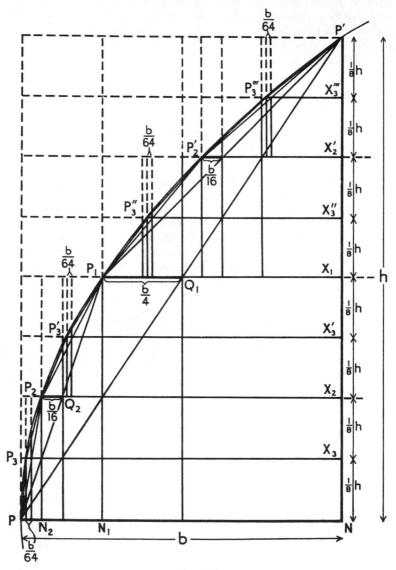

FIG. 54.

The area of the triangle PNP' has already been found. We now have to consider how we can find the area of the triangle PP$_1$P'. To do so we use this property of the parabola, that the *square* of the ordinate of a point is proportional to its abscissa. So, if we *halve* an ordinate, we *quarter* the corresponding abscissæ. In other words, since the ordinate of P$_1$ is half the ordinate of P', the abscissa of P$_1$ is a quarter of the abscissa of P'. Therefore PN$_1$ = $\frac{1}{4}b$. From the diagram we can see that Q$_1$X$_1$ = $\frac{1}{2}b$, so that

$$P_1Q_1 = PN - PN_1 - Q_1X_1$$
$$= b - \tfrac{1}{4}b - \tfrac{1}{2}b = \tfrac{1}{4}b.$$

We can now calculate the area of the triangle PP$_1$P'. We imagine it split into two triangles, each with its base on P$_1$Q$_1$. The bases of these triangles are both $\frac{1}{4}b$ and their heights are both $\frac{1}{2}h$, so that the area of the two together is

$$\tfrac{1}{4}b \times \tfrac{1}{2}h = \tfrac{1}{8}bh.$$

The next triangle to examine is PP$_1$P$_2$ and the calculation follows on the same lines we have already used for the triangle PP$_1$P'. The triangle PP$_1$P$_2$ is split into two triangles with their bases on P$_2$Q$_2$, the length of which we have to calculate. Using the property of the parabola connecting the ordinate and abscissa of any point, we notice that as P$_2$N$_2$ = $\frac{1}{2}$P$_1$N$_2$, it follows that PN$_2$ = $\frac{1}{4}$PN$_1$ = $\frac{1}{4} \times \frac{1}{4}b = \frac{1}{16}b$; by an argument similar to the one we used before it follows that P$_2$Q$_2$ is also $\frac{1}{16}b$. The area of the triangle PP$_1$P$_2$ is therefore

$$2 \times \tfrac{1}{2} \times \tfrac{1}{16}b \times \tfrac{1}{4}h = \tfrac{1}{64}bh.$$

By similar methods it can be shown that the triangle P$_1$P'P$_2$ has precisely the same area as triangle PP$_1$P$_2$ so that the area of these two triangles together is $\frac{1}{32}bh$.

We next calculate the area of the four triangles which have their vertices at P$_3$, P$_3$', P$_3$'', P$_3$'''. There is no need to repeat the arguments again ; the reader can convince himself, if he wishes, that each of these triangles can be split into two triangles each with a base of $\frac{1}{64}b$ and height $\frac{1}{8}h$. Their total area is therefore $8 \times \frac{1}{2} \times \frac{1}{64}b \times \frac{1}{8}h = \frac{1}{128}bh$.

Let us now summarise our results. The sum of the areas we have already calculated is

$$\tfrac{1}{2}bh + \tfrac{1}{8}bh + \tfrac{1}{32}bh + \tfrac{1}{128}bh = \tfrac{1}{2}bh(1 + \tfrac{1}{4} + \tfrac{1}{16} + \tfrac{1}{64}).$$

It is clear that the fractions inside the bracket follow a regular

sequence, each is $\frac{1}{4}$ of the previous fraction. We now say that were we to continue indefinitely the process of exhaustion in the same way, the total area would be the sum of the infinite series
$$\tfrac{1}{2}bh(1 + \tfrac{1}{4} + \tfrac{1}{16} + \tfrac{1}{64} + \ldots).$$
The summation formula for any infinite series of the kind
$1 + r + r^2 + r^3 \ldots$ where $|r| < 1$ is $\dfrac{1}{1-r}$; and in our example $r = \frac{1}{4}$. The area of the parabolic segment we have been considering is therefore
$$\tfrac{1}{2}bh \times \frac{1}{1 - \frac{1}{4}} = \tfrac{1}{2}bh \times \tfrac{4}{3} = \tfrac{2}{3}bh.$$

We should now double this area to include that part of the whole area below the axis of x.

We then finally arrive at the result that the area of the whole parabolic segment is $\frac{4}{3}bh$. If we were to draw the tangent at P, the vertex of the parabola and allow it to meet the diameters drawn through the ends of the chord which is the base of the segment, there would be formed a rectangle of base b and length $2h$, $i.e.$ with an area of $2bh$. As the area of the segment enclosed by the rectangle is $\frac{4}{3}bh$, we can see that the segment is exactly $\frac{2}{3}$ of the area of the rectangle.

Before we leave this excursion into classical Greek mathematics it should be noted that we have taken a specially simple case, that in which the chord of the segment is at right angles to the axis of the parabola. More generally it can be shown, as Archimedes did, that if the chord is inclined to the axis at any angle a similar relation holds. The enclosing rectangle becomes a parallelogram with one pair of sides parallel to the axis and the other pair consisting of the chord and a tangent to the parabola parallel to this chord. The area of the segment is still $\frac{2}{3}$ of the area of this parallelogram. We shall later find the area of this segment using the integral calculus ; we shall now more readily appreciate how powerful an instrument the calculus is.

CHAPTER XXX

SERIES

WE must now take a closer look at series of numbers which follow some regular sequence. The first type is that in which the numbers increase or decrease in equal steps by addition or subtraction of equal numbers. For example :—

$$1 + 3 + 5 + 7 + 9 + \ldots$$

or
$$500 + 496 + 492 + 488.$$

In the first of these examples the series increases from term to term by the addition of 2, and in the second it decreases by the subtraction of 4. Series of this kind are said to be " arithmetic."

If the terms of a series increase or decrease so that each successive term is obtained by multiplying or dividing its predecessor by a constant number, then we have what is called a " geometric " series. For example :—

$$1 + 3 + 9 + 27 + 81 \ldots \text{ etc. (multiplying by 3)}$$

or $1 - \frac{1}{4} + \frac{1}{16} - \frac{1}{64} \ldots$ (multiplying by $-\frac{1}{4}$ or dividing by -4).

There are general expressions for these two types of series. They are, for the arithmetic series,

$$a + (a + d) + (a + 2d) + (a + 3d) + \ldots (a + \overline{n - 1}d)$$

or $a + (a - d) + (a - 2d) + (a - 3d) + \ldots (a - \overline{n - 1}d)$

and for the geometric series,

$$a \pm ar + ar^2 \pm ar^3 \ldots + a(\pm r)^{n-1}$$

or
$$a \pm a \cdot \frac{1}{r} + a \cdot \frac{1}{r^2} \pm a \cdot \frac{1}{r^3} \ldots + a\left(\pm \frac{1}{r}\right)^{n-1}.$$

These particular types of series are sometimes called " progressions." The " a " is called the first term, the " d " is called the " common difference " of the arithmetic progression ; the " r," or $\frac{1}{r}$ is called the " common ratio " of the geometric progression.

We are particularly interested in knowing how to find the

sum of any number of terms in these series without going to the trouble of adding them term by term. For the arithmetic series the sum of " n " terms, called S_n, is found by writing the series down twice, once in the straightforward way, and once with the series reversed :—

$$S_n = a + (a + d) + (a + 2d) \ldots \quad \ldots \quad + (a + \overline{n - 1}d)$$
$$S_n = (a + \overline{n - 1}d) + (a + \overline{n - 2}d) + (a + \overline{n - 3}d) + \ldots + a$$

Adding these two series together term by term we have

$$2S_n = n \times (2a + \overline{n - 1}d)$$

or
$$S_n = \tfrac{1}{2}n\{2a + (n - 1)d\}.$$

An arithmetic progression always diverges so that we can find its sum only when n, the number of terms, is finite. We will use the formula to find the sum of the first 9 odd numbers—

$$1 + 3 + 5 + 7 + 9 + 11 + 13 + 15 + 17.$$

Here the first term, a, is 1, and the common difference, d, is 2, so that with $n = 9$

$$S_9 = \tfrac{1}{2} \cdot 9\{2 + 8 \times 2\} = \tfrac{9}{2} \times 18 = 81.$$

For geometrical progressions the summation formula is derived by another trick. The sum is written down twice again but in the second case the whole series is multiplied by r and then the second line is subtracted from the first :—

$$S_n = a + a \cdot r + a \cdot r^2 + \ldots + a \cdot r^{n-1}$$
$$rS_n = \quad a \cdot r + a \cdot r^2 \ldots \quad + a \cdot r^{n-1} + a \cdot r^n.$$

Subtracting,

$$S_n(1 - r) = a - ar^n$$

or
$$S_n = a\left(\frac{1 - r^n}{1 - r}\right) = a\left(\frac{r^n - 1}{r - 1}\right).$$

As an example of the use of this formula we will find the sum of the first six terms of the series

$$3 + 15 + 75 + 375 + 1875 + 9375.$$

Here $a = 3$, $r = 5$ and $n = 6$, so that

$$S_6 = 3\left(\frac{5^6 - 1}{5 - 1}\right) = 3 \cdot \frac{15625 - 1}{4} = 3 \cdot 3906 = 11{,}718,$$

a result which can be checked by direct addition.

We shall be more particularly concerned with the geometric

progression in which the common ratio, $| r |$, is a fraction less than 1. The importance of this kind of series is that it converges, that is, it steadily approaches a limit as the number of terms increases. We are therefore able to find its " sum to infinity." We have already shown that

$$S_n = \frac{a(1 - r^n)}{1 - r}.$$

In this formula $r^n \to 0$ as $n \to \infty$ as long as $| r | < 1$. The formula for the sum to infinity is therefore $S_\infty = \dfrac{a}{1 - r}$. We have already used this formula in the previous chapter to find the sum of the infinite series

$$1 + \tfrac{1}{4} + \tfrac{1}{16} + \tfrac{1}{64} \cdots$$

and have found it to be $\tfrac{4}{3}$. In a similar way the sum to infinity of the series

$$1 + \tfrac{1}{2} + \tfrac{1}{4} + \tfrac{1}{8} \cdots$$

is given by

$$S_\infty = \frac{1}{1 - r} = \frac{1}{1 - \tfrac{1}{2}}.$$
$$= 1 \div \tfrac{1}{2} = 2.$$

Again, for the series

$$\tfrac{1}{3} + \tfrac{1}{9} + \tfrac{1}{27} + \tfrac{1}{81} \cdots$$

$$S_\infty = \frac{\tfrac{1}{3}}{1 - \tfrac{1}{3}} = \tfrac{1}{3} \div \tfrac{2}{3} = \tfrac{1}{2}.$$

The most frequent case of this type of series is that in which $a = 1$. The series can then be written $1 + x + x^2 + x^3 + x^4 \cdots$, and if $| x | < 1$, we can write its sum as $\dfrac{1}{1 - x}$, a result which the reader can establish in a different way by expanding $\dfrac{1}{1 - x}$ or $(1 - x)^{-1}$ as a binomial expression.

From the examples it can be seen how, once a method of summing converging infinite series has been established, it is possible to solve many problems in the quadrature of plane surfaces bounded by curved lines. We have to find a method of " exhausting " the area of the figure in a way which produces a geometric series with descending terms. This method was successfully applied by Galileo and Viviani, the latter a friend of Leibniz. But Viviani at the height of his powers had the mortification of seeing the method he had used so well displaced by the much more powerful technique that is provided by the calculus. It is time now that we began to pluck the first fruits of the labours we have expended in mastering the groundwork that leads to the calculus.

CHAPTER XXXI

THE TECHNIQUE OF DIFFERENTIATION

WE are now prepared for a formal introduction to the differential calculus. We will begin by summarising the rules we have already established. If $y = f(x)$ is a function of x, then the differential coefficient of the function is

$$y' = \frac{dy}{dx} = Lt \frac{f(x + \Delta x) - f(x)}{\Delta x}$$

as $\Delta x \to 0$. In this equation Δx is a small increase in the value of x and $f(x + \Delta x)$ is the increase of y which is determined by the form of the function $f(x)$.

We now have to learn how to calculate the value of $\frac{dy}{dx}$ for various forms of $f(x)$; we begin by establishing a rule by which any power of x can be differentiated immediately, without going through a tedious algebraic operation. Differentiation of any function which is the sum of powers of x will then be no more difficult than division or multiplication.

We begin as usual with the simplest case—a function of the first degree in x, for example $y = x + b$.

Here $y + \Delta y = (x + \Delta x) + b$

and therefore

$$\frac{dy}{dx} = Lt \frac{(x + \Delta x) + b - (x + b)}{\Delta x}$$

$$= Lt \frac{x + \Delta x + b - x - b}{\Delta x}$$

$$= Lt \frac{\Delta x}{\Delta x} = 1.$$

It can be seen that the constant, b, does not affect the result ; it is cancelled out in the operation. The reason for this we shall understand later. Now for a function of the second degree, let us say $y = x^2 - 7$. For this function

$$\frac{dy}{dx} = \underset{t}{L} \frac{(x + \Delta x)^2 - 7 - (x^2 - 7)}{\Delta x}$$

$$= \underset{t}{L} \frac{x^2 + 2x \cdot \Delta x + (\Delta x)^2 - 7 - x^2 + 7}{\Delta x}.$$

$$= \underset{t}{L} \frac{2x \cdot \Delta x + (\Delta x)^2}{\Delta x}$$

$$= \underset{t}{L} (2x + \Delta x)$$

$$= 2x.$$

Once again the constant disappears, as we might have expected, so we shall not bother to include one in the next example. We could go on now and differentiate functions of the third, fourth and higher degrees until the rule becomes quite clearly apparent ; this would be to use the " Method of Induction." It would be unnecessarily laborious however because we are familiar with and can make use of the Binomial Theorem. This allows us to jump at once to the general case, that is to the function $\quad y = x^n.$

The differentiation of x^n proceeds in the usual way.

$$\frac{dy}{dx} = \underset{t}{L} \frac{(x + \Delta x)^n - x^n}{\Delta x}.$$

Now the binomial $(x + \Delta x)^n$ is expanded so that

$$\frac{dy}{dx} = \underset{t}{L} \frac{\{x^n + {}_nC_1 \cdot x^{n-1} \cdot \Delta x + {}_nC_2 \cdot x^{n-2} \cdot (\Delta x)^2 + \ldots\} - x^n}{\Delta \mathbf{x}}$$

The two terms x^n cancel out leaving all the remaining terms as multiples of Δx. We can therefore divide each of these terms by the Δx of the denominator :—

$$\frac{dy}{dx} = \underset{t}{L} \{{}_nC_1 \cdot x^{n-1} + {}_nC_2 \cdot x^{n-2} \cdot \Delta x + {}_nC_3 \cdot x^{n-3}(\Delta x)^2 \ldots \}.$$

In this expression the second term is of the first order in Δx, the third term is of the second order, and so on. As $\Delta x \to 0$, all these terms therefore disappear, leaving only the first ; so we can write $\quad \dfrac{dy}{dx} = {}_nC_1 \cdot x^{n-1} = n \cdot x^{n-1}.$

This then is the general rule for differentiating powers of x. Let us test it immediately on functions we have already differentiated from " first principles." If $y = x$, the unwritten index is 1, so that $n = 1$ and $y' = 1 \cdot x^{1-1} = 1 \cdot x^0 = 1 \cdot 1 = 1$. If $y = x^2$ then $n = 2$ and $y' = 2 \cdot x^{2-1} = 2 \cdot x^1 = 2x$. Finally, if $y = a$, a constant, then $y = ax^0$ since $x^0 = 1$, and $y' = a \cdot 0 \cdot x^{0-1} = a \cdot 0 \cdot x^{-1} = 0$. All these results accord with our previous calculations. For any other value of x the rule applies just as simply, for if $y = x^{16}$, $\dfrac{dy}{dx} = y' = 16 \cdot x^{15}$.

Notice that when a power of x is differentiated the power is reduced by 1. The rule is valid not only for integral values of n ; it is equally true if n is negative or fractional. If $y = \sqrt{x}$, expressing this in an index form,

$$y = x^{\frac{1}{2}} \text{ and } \frac{dy}{dx} = \tfrac{1}{2} \cdot x^{\frac{1}{2}-1} = \tfrac{1}{2} \cdot x^{-\frac{1}{2}} = \tfrac{1}{2} \cdot \frac{1}{x^{\frac{1}{2}}}.$$

$$= \frac{1}{2\sqrt{x}}.$$

If $y = x^{-3}$, then

$$\frac{dy}{dx} = -3 \cdot x^{-3-1} = -3 \cdot x^{-4} \text{ or } \frac{-3}{x^4}.$$

The observant reader will have noticed that our rules cover only those cases in which the coefficient of the power of x is 1. What happens if we have to differentiate the function $y = 3x^2$, where we have a coefficient which is not unity ? The rule for this case is that this constant remains as a multiplying factor in the result. The rule can be verified by differentiating $3x^2$ by the first method :—

$$\text{If } y = 3x^2, \frac{dy}{dx} = L_t \frac{3(x + \Delta x)^2 - 3x^2}{\Delta x}$$

$$= L_t \frac{3[x^2 + 2x \cdot \Delta x + (\Delta x)^2] - 3x^2}{\Delta x}$$

$$= L_t \frac{3x^2 + 6x \cdot \Delta x + 3(\Delta x)^2 - 3x^2}{\Delta x}$$

$$= L_t (6x + 3\Delta x) = 6x.$$

Summarising the effect of differentiating functions of powers of x which also contain constants, we can say that constants of addition (or subtraction) disappear, constants of multiplication (or division) remain.

The next step is to consider the most general form of $f(x)$ when this function consists of the sum of different powers of x, *i.e.*,
$$y = ax^n + bx^{n-1} + cx^{n-2}. \ldots$$
We will take as an example
$$y = 4x^3 - 7x^2 + 9x - 26.$$
If this function is differentiated from first principles the result will eventually be found to be
$$\frac{dy}{dx} = 12x^2 - 14x + 9.$$

The same result can be obtained by applying the rules we have established to each of the four terms in turn. Differentiating $4x^3$ we have $4 . 3x^2 = 12x^2$; for $- 7x^2$ we have $- 7 . 2x = - 14x$; for $9x$ we have $9 . 1 = 9$; and finally the unattached constant, $- 26$, disappears. In other words, the differential coefficient of an algebraic sum of terms is the algebraic sum of the differential coefficients of the several terms.

At this point it is necessary to give an emphatic warning that the rules for differentiation which we have stated do not apply to functions which are themselves products or quotients of two or more functions, *e.g.*,
$$y = (x^2 + 3x + 1)(5x - 16), \quad y = \frac{2x^3 - 7}{3x + 9}.$$

For such cases special developments of our rules have to be devised and they are beyond the scope of this book. If the functions can be multiplied out to give a sum of powers of x, then this sum can of course be differentiated by our rules, but this rarely applies to quotients although it may apply to products.

Having shown how to calculate the differential coefficient of $f(x)$ when this represents a " polynomial in x," *i.e.*, a sum of powers in x with constant coefficients, we must now find a geometrical meaning for it. The differential coefficient has a wide field of application ; some of this will be described in the following chapters. Let us first summarise what we know. The differential coefficient is the limiting value of the ratio $\Delta y : \Delta x$ when these become infinitesimally small, Δx being a small increment in x and Δy being the corresponding increment in y determined by the relation $y = f(x)$. What does this mean when we consider the graph of the curve $y = f(x)$?

As the " element " Δs is known to be a piece of the tangent to the curve, we can continue or " produce " it so that it cuts the x-axis (Fig. 55). The angle between the tangent and the

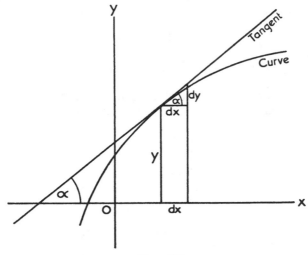

positive direction of the x-axis is equal to the angle between the elements Δs and Δx (since Δx is parallel to the x-axis) when these elements become infinitesimally small. When this occurs $\dfrac{\Delta y}{\Delta x} \rightarrow \dfrac{dy}{dx}$, but at the same time $\dfrac{\Delta y}{\Delta x}$ is, from the standpoint of trigonometry, the tangent of the angle α in the triangle. Therefore

$$\frac{dy}{dx} = \tan \alpha$$

where α is now also the slope of the tangent.

If then we know the equation of a curve as $y = f(x)$, we can find the value of y for any particular numerical value of x by substituting this value in $f(x)$ and we can therefore mark a point on the curve for this value of x. From the value of $y' = f'(x)$ we can also find, by substituting the same value of x in $f'(x)$, the numerical value of the slope of the curve at that point. In fact it enables us to draw the tangent at any point of a curve, given the equation of the curve, without first drawing the curve. Let us take the parabola $y = \dfrac{x^2}{10} + 3$ as an example and find the tangent to it at the point where $x = 4$.

When $x = 4$, $y = \frac{16}{10} + 3 = 4\cdot6$. The point P has the co-ordinates $x = 4$, $y = 4\cdot6$. Differentiating the function,

$$\frac{dy}{dx} = \tfrac{1}{10} \cdot 2x = \frac{x}{5}.$$

When $x = 4$, $\dfrac{dy}{dx} = 0\cdot8$ and therefore $\tan \alpha = 0\cdot8$. By referring to a table of natural tangents we see that α must be $38°\,40'$. We can now draw it, as we have done in Fig. 56. If

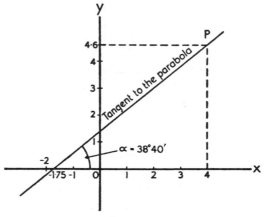

FIG. 56.

we were now to draw the parabola itself we should find that it does in fact pass through the point P, just touching the line whose slope we have calculated.

CHAPTER XXXII

MAXIMA AND MINIMA

THE discovery of the differential coefficient arose from the investigation of the problems concerned with tangency, that is the analytical condition that lines and curves should touch each other. These problems were examined and results obtained with increasing exactitude during the seventeenth century from the time of Descartes. They were very important to mathematicians for the following reasons. If any two physical quantities are related by a law that can be expressed in mathematical terms, then this relation can be represented by a function and therefore by a graph. Graphs frequently show " ups and downs," and these are often the most important characteristics of the curve. The highest and lowest points of the curve, sometimes called its " turning values," are known as its " maxima and minima "—terms which may be familiar to the non-mathematician. At these turning points the tangent to the curve is parallel to the x-axis, as can be seen from Fig. 57. This important fact must now be related to our knowledge of the differential coefficient. Since the latter is a measure of the slope of the curve, and since the

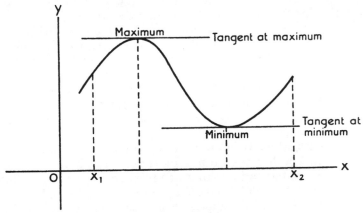

FIG. 57.

slope of the curve at these points is zero, the differential coefficient must also be zero for these particular points. Analytically expressed, this is :—

$$\tan \alpha = \frac{dy}{dx}$$

$$\alpha = 0, \quad \therefore \ \tan \alpha = 0$$

$$\therefore \ \frac{dy}{dx} = 0.$$

Hitherto we have always found the numerical value of $\frac{dy}{dx}$ at a particular point by substituting the value of the abscissa of the point in the function $f'(x)$. Now we work in reverse. We know the value of $f'(x)$, namely 0, but we do not know the values of x for which this is true. We therefore write $f'(x) = 0$ and solve this as an equation in x. When the values of x which are the roots of this equation are substituted in $f(x)$ we get the ordinates of the corresponding points, *i.e.*, of the points at which the maxima and minima occur.

Here an ambiguity arises. We have shown how to find the maxima and minima, but how do we know which is which ? There is a means of discriminating between them but we do not propose to describe it here. In most practical cases there is little need for further analysis. The conditions of the problem, or the graph of the curve, usually provide sufficient information for us to decide which case in fact our solution may represent. We are now ready to tackle the problem and will do so by means of a few interesting examples.

First, here is a classical example. A flat piece of tin-plate, square in shape and with sides " a " inches long, is to be turned into a square-based box open at the top. In order to do this, four smaller squares must be cut out of the corners of the plate so that the four sides can be folded up and soldered together (Fig. 58). The problem is to find out how big the cut-out squares must be to give the box its maximum cubic capacity.

It is clear that as we cut out corner squares of different sizes, so we shall get boxes of different shapes and sizes. At one extreme, if we cut out nothing at all, there is nothing to fold up and we have a flat " box " of no capacity. At the other extreme, if we cut away squares whose sides are $\frac{1}{2}$ " a " inches, we are left with a box that has no base. Somewhere in between these values, that is between 0 and $\frac{1}{2}a$, there is an unknown value of the side of the corner squares, we will call it x inches, which is the value we are seeking. How do we find it ?

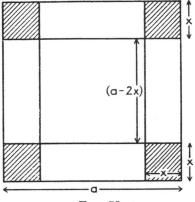

FIG. 58.

One method is to allow x to take a range of values between 0 and $\frac{1}{2}a$, and for each value to calculate the capacity of the box that would thus be formed. A graph could then be drawn with values of x as abscissæ and the corresponding capacities as ordinates. From the graph the highest value reached could then be found. This method is laborious and in the end gives only an approximate answer.

It would be quicker of course to calculate the capacity of the box for the general value of x. We should then obtain a general formula for the capacity of the box. This formula could then be treated as a function of x and so be plotted as a curve of which the highest point is estimated as before ; but the same objections to this method could be made as to the first suggested method.

The method using the calculus provides the exact result and the following steps have to be carried out :—

1. Find out the " range " of the function, *i.e.*, the limits between which the maximum is required ; in this case we know that the required value lies between 0 and $\frac{1}{2}a$.

2. Find the function with which the problem is concerned ; in this case that is, express the capacity of the box in terms of the unknown, x, and any constants which may be known, *e.g.*, the side of the square, a inches. The capacity is given by the

formula (Area of base) $\times$ Height.   The base of the box will be a square of side $(a - 2x)$ inches ;   its height will be x inches. We therefore have the formula :—

$$\text{Capacity} = (a - 2x)^2 \, . \, x \text{ cubic inches.}$$

Hence the function we have to consider is

$$
\begin{aligned}
y &= (a - 2x)^2 \, . \, x \\
  &= (a^2 - 4ax + 4x^2)x \\
  &= 4x^3 - 4ax^2 + a^2x.
\end{aligned}
$$

3. Find the differential coefficient of the function.

Here

$$
\begin{aligned}
\frac{dy}{dx} &= 4 \, . \, 3x^2 - 4a \, . \, 2x - a^2 \, . \, 1 \\
              &= 12x^2 - 8ax + a^2.
\end{aligned}
$$

4. Put this differential coefficient equal to 0 and solve the resulting equation for x.   This is the analytical way of saying that at the maximum point the slope of the tangent must be zero, $i.e.$, the tangent is parallel to the x-axis.   In our example we have

$$12x^2 - 8ax + a^2 = 0.$$

This is a quadratic equation for x.   The constant " a " is treated just as though it were a concrete number so that, after dividing this equation by 12 to reduce the coefficient in x^2 to 1, we can use the formula we derived in Chapter XXIII.   The equation is now

$$x^2 - \frac{2a \, . \, x}{3} + \frac{a^2}{12} = 0,$$

of which the solution, by formula, is

$$
\begin{aligned}
x &= \frac{2a}{6} \pm \sqrt{\frac{4a^2}{36} - \frac{a^2}{12}} \\
  &= \frac{2a}{6} \pm \sqrt{\frac{4a^2 - 3a^2}{36}} \\
  &= \frac{2a}{6} \pm \sqrt{\frac{a^2}{36}} \\
  &= \frac{2a}{6} \pm \frac{a}{6}.
\end{aligned}
$$

Hence

$$x = \frac{a}{6}, \text{ or } \frac{3a}{6} = \frac{a}{2}.$$

At these values of x, this is a maximum or minimum value

of y ; in other words, corner squares of these sizes will give us maximum or minimum values of the capacity.

5. Now we have to decide which of these values is the one we want. In this and all the other examples we shall give we shall be able to decide from our knowledge of the practical problem. Here we know already that if $x = \dfrac{a}{2}$ the box will have no base, so this solution can be rejected without further consideration. It follows that the solution we need is that $x = \frac{1}{6} \cdot a$; the side of the corner squares should be $\frac{1}{6} a$ inches, giving a box of side $(a - \frac{2}{6}a)$ inches $= \frac{2}{3}a$ inches.

To take a numerical case, let us suppose that the original square of tinplate had a side of 60 inches. Then for a box of maximum capacity the corner squares cut out should have sides of 10 inches. The box would then have the dimensions : length 40 inches, breadth 40 inches, height 10 inches, with a capacity of 16,000 cubic inches. Suppose we try another value of x, say 5 inches. The dimensions of the box would then be : breadth and length, each 50 inches, height 5 inches, capacity 12,500 cubic inches. Suppose lastly that we cut out squares of side greater than 10 inches, say 15 inches. Then the dimensions of the box would be $30 \times 30 \times 15$ inches and its capacity 13,500 cubic inches. It will be found that whatever other values of x we try, none will succeed in improving on the value the calculus method has given us.

With the experience of working out this example behind us, a second example drawn from statics will present few difficulties.

From a tree trunk of circular cross section and given length, a rectangular beam of maximum strength is to be cut (see Fig. 59). The radius of the tree trunk is known : we shall call it r. The strength of a beam depends on its height (or depth) and its breadth and for our purposes it can be represented by the formula $F = h^2 b$, in which F stands for strength, h is the depth of the beam and b is its breadth. In Fig. 59 two of the infinitely many possible beams are shown. We will call the breadth of the beam $2x$. As there is a right-angled triangle we can apply the theorem of Pythagoras in order to provide a relation between r, x and h. It is

$$\left(\frac{h}{2}\right)^2 = r^2 - x^2$$

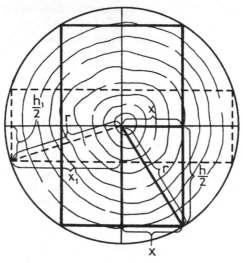

Fig. 59.

or $$\frac{h^2}{4} = r^2 - x^2$$

or $$h^2 = 4(r^2 - x^2).$$

Now it is the strength, F, which we wish to make a maximum, so we express F in terms of x.

$$F = h^2 \,.\, b = 4(r^2 - x^2) \,.\, 2x.$$

The function which we are to consider is therefore

$$y = 8r^2x - 8x^3.$$

What is the " range " of the function ? The breadth, $2x$, can lie between 0 and $2r$. In the first case it has no breadth at all, and in the second case the depth is zero. It is also to be noticed that the breadth must be a positive number. A beam of negative breadth is as useless to an engineer as a beam of no depth.

We now differentiate the function :—

$$\frac{dy}{dx} = 8r^2 \,.\, 1 - 8 \,.\, 3x^2$$

$$= 8r^2 - 24x^2.$$

Next we put the differential coefficient equal to 0 and solve the resulting equation for x.

$$8r^2 - 24x^2 = 0$$
$$24x^2 = 8r^2$$
$$x^2 = \tfrac{1}{3}r^2$$
$$x = \pm \frac{1}{\sqrt{3}} r = \pm \frac{\sqrt{3}}{3} \cdot r$$

We are interested only in the positive sign. The strongest beam is therefore that for which the breadth is $\dfrac{2\sqrt{3}}{3} \cdot r = 1\cdot 15 \cdot r$.

We will close this chapter by finding a minimum. Inside a given square it is possible to inscribe an infinity of smaller squares like that shown in Fig. 60. Which of these is the smallest ?

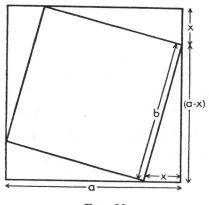

Fig. 60.

When the length x marked in the diagram is equal to 0 we obviously have the biggest possible inscribed square. If x increases, the inscribed square will become smaller, but after some point which we want to find it is evident that the square will again increase as x approaches the length a. The range of x is therefore 0 to a. Let us take the side of the inscribed square to be b inches. Then the area of the square will be b^2 square inches—and this is the quantity which has to be made a minimum. Hence $y = b^2$. We now have to connect b with x ; once again Pythagoras comes to the rescue, since

$$b^2 = x^2 + (a - x)^2$$
$$= x^2 + a^2 - 2ax + x^2$$
$$= a^2 - 2ax + 2x^2.$$

Here is the function, $y = a^2 - 2ax + 2x^2$, and differentiating,

$$\frac{dy}{dx} = -2a \,.\, 1 + 2 \,.\, 2x$$
$$= 4x - 2a.$$

At a minimum or maximum $\dfrac{dy}{dx} = 0$ so that we put $4x - 2a = 0$ and solve this equation for x.

$$4x = 2a$$
$$x = \tfrac{1}{2}a.$$

We have already noticed that maximum values for the area of an inscribed square occur at the limits of the range, *i.e.*, where $x = 0$ and $x = a$, and that the area decreased as x moved away from these points. If there is only one turning point we should therefore expect it to be a minimum. It is not surprising that in this symmetrical system we should find the minimum occurs at the middle point. As a check suppose the large square has a side of 12 inches. Then when $x = \dfrac{a}{2}$, the area of the inscribed square is 72 square inches, and when $x = \dfrac{a}{4}$ or $\dfrac{3a}{4}$, the area is 90 square inches in both cases, as the reader can verify from the formula $b^2 = a^2 - 2ax + 2x^2$.

CHAPTER XXXIII

THE TECHNIQUE OF INTEGRATION

WE must take up the thread of our progress again towards an understanding of integration. We have seen that the function under the integral sign is the differential coefficient of the integrated function. Then if $y = f(x)$, and $y' = f'(x)$, then $\int y'dx$ or $\int f'(x)dx = f(x)$.

Meanwhile we have learnt the basic rules of differentiation and can use them to find basic rules of integration. We shall differentiate a function and then by integration try to restore the differential coefficient to its original state. This will teach us much about the nature of the integration process. Then when we have learnt how to apply integration we shall have completed the task we set out to do in this book.

We will choose as the function to be differentiated

$$f(x) = y = 2x^3 - 7x^2 + x + 89$$

from which

$$\frac{dy}{dx} = f'(x) = y' = 2 \cdot 3x^2 - 7 \cdot 2x + 1 \cdot x^0$$
$$= 6x^2 - 14x + 1.$$

It is already noticeable that there is a manifold ambiguity in integration. The constant term 89 has disappeared in the course of differentiation and no one can see how to put it back. A function with any other constant term 88, 90, 2, $- 6$ or even 0 would have produced the same differential coefficient, so how do we decide which of these to put back when we reverse the process? We overcome the difficulty in general terms by writing

$$\int f'(x)dx = f(x) + c$$

where c stands for " any constant." Sometimes the term c is omitted but in such cases it is always implied, and the integral is said to be " indefinite." In actual calculation we use " definite " integrals and the doubt about c does not arise, as we shall see later. However, it is important that we shall understand the geometrical explanation of the indeterminate value of the constant c.

To return to our concrete example, since we know that $\int f'(x)dx = f(x) + c$, we can write

$$\int(6x^2 - 14x + 1)dx = 2x^3 - 7x^2 + x + c.$$

For the present we will ignore the term c and compare the two sides of this equation term by term :—

$$f(x) = 2x^3 - 7x^2 + x$$
$$f'(x) = 6x^2 - 14x + 1.$$

The first point to note is that, just as in differentiation, we can integrate term by term. Applying this rule immediately,

$$\int(6x^2 - 14x + 1)dx = \int 6x^2 dx - \int 14x dx + \int 1 dx.$$

The integral of a sum of powers of x is the sum of integrals of the powers of x taken separately. The coefficients can also be treated as constants of multiplication ; so that we can write

$$\int(6x^2 - 14x + 1)dx = 6\int x^2 dx - 14\int x dx + 1\int 1 dx.$$

Now we will look at the powers of x. From $6x^2$ we have to get $2x^3$, from $14x$ we must get $7x^2$ and from 1 we must get x. The first point to notice is that the power of x is raised by 1 on integration, as we might expect if we remember that differentiation reduced the power by 1. Generally, therefore, $\int x^m dx$ will give a term $= x^{m+1}$, but the coefficient still has to be found. Returning to our example, we see that x comes from 1 ; $7x^2$ from $14x$ (we have to divide by 2) ; $2x^3$ from $6x^2$ (we have to divide by 3). All these requirements will be met if we make the general rule $\int x^m dx = \dfrac{1}{m+1} \cdot x^{m+1}$. For

$$\int 1 dx = \int x^0 dx = \frac{1}{0+1} \cdot x^{0+1} = x \ ;$$

$$\int 14x dx = 14\int x^1 dx = 14 \cdot \frac{1}{1+1} x^{1+1} = \frac{14}{2}x^2 = 7x^2,$$

and $$\int 6x^2 dx = 6\int x^2 dx = 6 \cdot \frac{1}{2+1} x^{2+1} = \frac{6}{3}x^3 = 2x^3.$$

Here is the secret of the integral sign before us and it can be seen, as we promised, that it is no more difficult a command to obey than any other mathematical command we have met, though our present rules allow us to apply it only to " polynomials " of x, that is, sums of powers of x with constant coefficients. We will not conceal the fact that we may not yet be able to integrate more complicated functions, as for example the one which arises in the rectification of the parabola :—

$$\int_0^x \sqrt{1 + \frac{x}{k^2}}\, dx = \frac{x}{2}\sqrt{1 + \frac{x^2}{k^2}} + \frac{k}{2}\log\left(\frac{x}{k} + \sqrt{1 + \frac{x^2}{k^2}}\right).$$

There are many more tricks to learn about integration than we can hope to deal with in this book. Many of the simpler functions which arise in integration problems can be found with their integrals in special tables, but the integration of complicated functions is an art in itself.

There are in fact some functions, and they sometimes look very simple ones, which have no integrals in the ordinary sense. In other words, we can write down functions which are not the derivatives of any existing functions, and some of these are very important in higher mathematics. In such cases, however, it is always possible to evaluate them to an approximation of as close a degree as we require.

Having stressed the limitations of our present knowledge of integrals, we must not be discouraged. There are many applications of the integral calculus now possible for us and, with the main principles established, we can solve many problems hitherto insoluble.

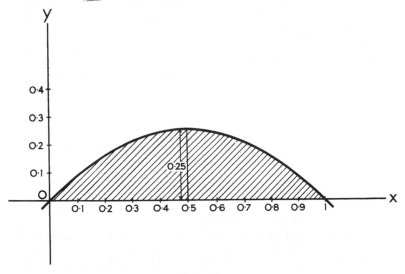

Fig. 61.

We will return to our problem of quadrature. In Fig. 61 part of the curve is drawn whose equation is

$$y = x - x^2.$$

We shall concern ourselves only with that part of the curve which lies in the first quadrant. As it can be seen, $y = 0$ when $x = 0$ and when $x = 1$, so that the curve must cut the x-axis where $x = 0$ and $x = 1$. Our problem is to find the area enclosed by the curve and the x-axis.

We must not forget that we can now find maxima and minima and here is an example on which we can practise. The range of x in which we are interested is from $x = 0$ to $x = 1$. Since

$$y = x - x^2$$
$$\frac{dy}{dx} = 1 - 2x.$$

We put $\dfrac{dy}{dx} = 0$ and solve for x :—

$$1 - 2x = 0$$
or $$x = \tfrac{1}{2}.$$

The maximum (or minimum) therefore occurs at the point where $x = \tfrac{1}{2}$. The ordinate at this point is $y = x - x^2 = \tfrac{1}{2} - \tfrac{1}{4} = \tfrac{1}{4}$. The highest point of the curve above the x-axis is therefore $\tfrac{1}{4}$, and it is at the point halfway between $x = 0$ and $x = 1$. This is clearly its " maximum."

Now for its area. We know that it can be expressed as a " definite " integral, that is with upper and lower limits to express the range of the integration. Thus in our example we have to find

$$\int_0^1 (x - x^2)dx.$$

How do we handle this definite integral ? First we treat it as though it were an indefinite integral, applying the rules of integration as we have learnt them :—

$$\int (x - x^2)dx = \int x\,dx - \int x^2 dx$$
$$= \tfrac{1}{2}x^2 - \tfrac{1}{3}x^3.$$

(The constant is omitted for the present.)

Secondly, we deal with the limits. The rule is—substitute the values of the limits in the integral in turn and then subtract the integral with the lower limit from that with the upper limit. In the general case we have $\int f'(x)dx = f(x) + c$, an indefinite integral. If we now put in the limits $x = a$ and $x = b$, if we " integrate from a to b," we have

$$\int_a^b f'(x) \,.\, dx = [f(x) + c]_a^b$$

which by the rule just stated becomes

$$f(b) + c - f(a) - c = f(b) - f(a).$$

So, in a definite integral the value of c is not required because it cancels out in any case. In the numerical example we have before us the limits are 0 and 1. The indefinite integral was $\frac{1}{2}x^2 - \frac{1}{3}x^3 + c$. If we now substitute the limit values we have

$$[\tfrac{1}{2} \,.\, 1^2 - \tfrac{1}{3} \,.\, 1^3 + c] - [\tfrac{1}{2} \,.\, 0^2 - \tfrac{1}{3} \,.\, 0^3 + c]$$
$$= \tfrac{1}{2} - \tfrac{1}{3} + c - c = \tfrac{1}{6}.$$

The quadrature is complete. The area under the curve is $\frac{1}{6}$ square units, the units being those (equal) units in which x and y are measured.

CHAPTER XXXIV

MEAN VALUE AND DEFINITE INTEGRALS

WE will use the last example to illustrate another point. The base line of the figure is 1 unit and its area is $\frac{1}{6}$ square units. It therefore has the same area as a rectangle of base 1 unit and height $\frac{1}{6}$ unit. We can look upon the figure as consisting of innumerable ordinates, some greater than $\frac{1}{6}$ units, some less ; but $\frac{1}{6}$ unit is the middle, the average or mean value of the ordinates. The " mean value " is merely the " average " of everyday speech, on which we need not dilate. The general formula for finding the mean (or average) of the " n " quantities represented by $a_1, a_2, a_3 \ldots a_n$ is :—

$$\text{Mean} = \frac{a_1 + a_2 + a_3 \ldots a_n}{n}.$$

In our example, however, the number of ordinates is not finite. If $y_1, y_2, y_3 \ldots$ are the ordinates it would not be possible to calculate the mean arithmetically from the formula

$$\frac{y_1 + y_2 + y_3 \ldots y_\infty}{\infty}.$$

If we imagine the ordinates spaced out at intervals of Δx along the x-axis then the number of ordinates would be $\dfrac{b - a}{\Delta x}$ where $x = b$ and $x = a$ are the limits of the range of x. Then the mean value of the ordinates is equal to their sum divided by $\dfrac{b - a}{\Delta x}$ or

$$\text{Mean} = \sum_a^b y \times \frac{\Delta x}{b - a}.$$

As $\Delta x \rightarrow 0$ this becomes

$$\frac{\int_a^b y\,dx}{b - a} = \frac{\int_a^b f(x)\,dx}{b - a}.$$

We will now check this formula for our example. The integral $\int_0^1 (x - x^2)\,dx = \frac{1}{6}$, $b - a = 1 - 0 = 1$, and so $\frac{1}{6}$ is the mean value of the ordinates.

We have all seen those instruments like barographs which automatically record their readings graphically on paper wrapped round a circular drum which is rotated by clockwork. As the pen which records the readings keeps in contact with the steadily moving drum, the graph drawn is a continuous, and therefore differentiable, function. The curves drawn by the instruments are, however, so irregular that it is impracticable to represent them by formulæ.

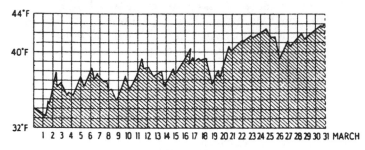

FIG. 62.

An example is given in Fig. 62 of a temperature recording for the month of March. We are asked to find the mean (or average) temperature for the month ; in other words, we have to find the mean of all the ordinates. Since the equation of the curve is unknown, we cannot *calculate* the area under the curve but we can find a close approximation to it either by counting the squares or by comparing the weight of the cut-out figure with that of a unit square. When we have done this, then we have the equivalent of the integral $\int_a^b f(x)dx$. The denominator $b - a = 31$ in our particular case, so that when we know the area in suitable units we can estimate the mean value of the temperature for the month. Here we have had no need to integrate at all ; but the method depends for its justification on the methods of the calculus, as we cannot find the middle value of an infinity of ordinates without using the infinitesimal.

Before we tackle another quadrature we should consider once again the question of the constants that arise in integration. We have already seen in the previous chapter how they cancel out when definite integrals are used. If $\int f'(x)dx =$

$f(x) + c$, then the definite integral

$$\int_a^b f'(x)dx = [f(b) + c] - [f(a) + c]$$
$$= f(b) - f(a).$$

Geometrically this difference is merely the taking away of one area from another, though of course only for the final result. From a different point of view the indeterminate integral may be looked upon as a general formula for the area under the curve, which only gives a concrete value when the values of the extreme abscissæ are substituted.

We must now consider what are known as "integral" curves. We know that every function can be represented by a curve. If we integrate a function as, for example,

$$\int(x - x^2)dx = \frac{x^2}{2} - \frac{x^3}{3} + c,$$

we have a new function and therefore presumably a new curve, an "integral" curve. Unfortunately we do not know the value of c. It can be positive, negative or zero. What is the meaning of this constant ? The answer to this question is, that whatever value we give the constant, the shape of the curve remains the same ; all that happens is that the curve is pushed up or down the y-axis. In Fig. 63 a number of parabolas with the equation $y = \frac{x^2}{4} \pm c$ are drawn.

Generally speaking. as the constant can be given any value between $+ \infty$ and $- \infty$, the indeterminate integral is represented by a whole "family" of similar curves so close to each other that together they fill the whole plane. This characteristic is of great importance in physics. Many processes can be represented by "differential" equations which describe analytically the general features of the process, e.g., of wave motion. These equations, solved by integration, have generally an infinity of solutions. Any particular solution required is determined by the constants and the limiting conditions associated with the particular problem—but this is by the way.

Suppose we have to integrate the function $y' = \frac{x}{2}$. We obtain $y = \int \frac{x}{2} dx = \frac{x^2}{4} + c$. The definite integral

$$\int_a^b \frac{x}{2} dx = \frac{b^2}{4} + c - \frac{a^2}{4} - c = \tfrac{1}{4}(b^2 - a^2).$$

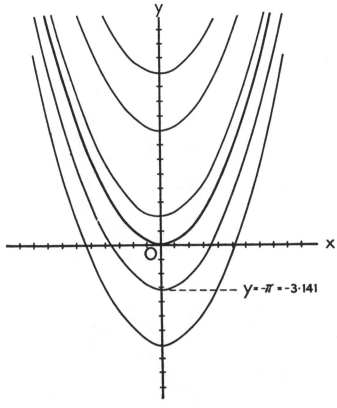

$$y = -\pi = -3\cdot 141$$

Fig. 63.

The ordinates of the indefinite integral curve where $x = b$ and $x = a$ are respectively $\frac{b^2}{4} + c$ and $\frac{a^2}{4} + c$. Hence the definite integral with the limits a and b is equal to the difference of these two ordinates which correspond to the limits of the integration. This point is illustrated in Fig. 64. OPP^1 is the graph of the function $y = \frac{x}{2}$, a straight line which passes through 0. The curves are integral curves of this function, namely $y = \frac{x^2}{4} + c$, with different values given to c.

The area to be calculated is shaded in the diagram ; it is

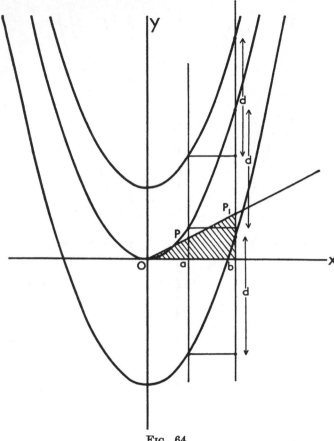

FIG. 64.

the difference of the triangles OP_1b and OPa. The lengths marked " d " are in every case the differences of the ordinates at $x = b$ and $x = a$ for each of the curves. This length d is always proportional to the shaded area wherever the points a and b may be (on the positive side of the x-axis).

We have already mentioned differential curves. If we begin with the function $x - x^2$, its integral is $\dfrac{x^2}{2} - \dfrac{x^3}{3} + c$ and its differential coefficient is $1 - 2x$. All three functions can of course be represented by curves and the three curves are shown in Fig. 65, c being equal to $+ 1$ in the integral curve.

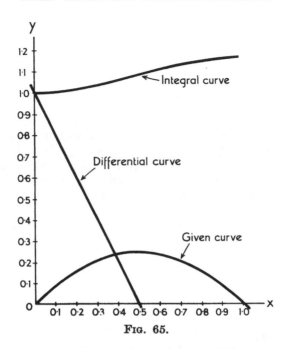

FIG. 65.

They can be plotted by the reader by giving x different convenient values, 0, 1, 2, etc., and marking the three points corresponding to each value of x.

It is worth noticing that the " differential " curve of the " integral " curve $y = \dfrac{x^2}{2} - \dfrac{x^3}{3} + 1$ is of course $y = x - x^2$, the original curve. Also the integral curves of $y = 1 - 2x$, namely $y = x - x^2 + c$, include the original curve for which case $c = 1$.

The relations between an x-function, its " integral " and its " differential " are put to practical use in the designing of mechanisms in which it may be important to know the velocity and the acceleration of moving points at different instants during a cycle.

CHAPTER XXXV

FURTHER PROBLEMS OF AREA

WE have now established the basic principles of the calculus, and though we cannot delve into the subject more deeply we can test our system of hard-won knowledge in a few simple cases. Let us begin by using the calculus to " square the square," if we may be permitted such an expression. For this case the curve under which the square lies will be a straight line parallel to the x-axis.

If we decide to call the length of the side of the square " a," then the equation of the " curve " will be $y = a$ (Fig. 66).

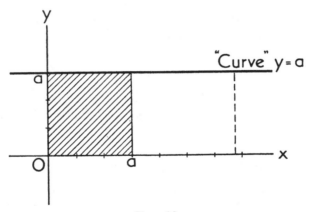

FIG. 66.

We can write this equation as $y = ax^0$. The limits of the integration which we now have to perform (indicated by the shaded area in the diagram), are 0 and a. Hence

$$\text{Area} = \int_0^a ax^0 dx = a\int_0^a x^0 dx = a[x]_0^a$$
$$= a(a - 0) = a^2.$$

We have found the area of the square by integration! To find the area of a rectangle with sides a and b, we carry out the

236

same integration but with different limits, namely 0 to b. We therefore have

$$\text{Area} = \int_0^b ax^0 dx = a\big[x\big]_0^b = a(b - 0)$$
$$= ab.$$

This result is correct too—as we know.

Now let us turn our attention to Archimedes' problem, the quadrature of the parabola. He obtained his result (Chapter XXIX) by finding the area of a series of inscribed triangles which led to the infinite series $1 + \frac{1}{4} + \frac{1}{16} + \cdots$ We must begin with an equation. In text-books the equation of the parabola is usually given as $y^2 = 4px$. We can simplify this formula a little to make our task easier by putting $p = \frac{1}{4}$. The equation then becomes $y^2 = x$ or $y = \sqrt{x}$. We can do still more. The parabola $y^2 = x$ has its axis lying along the x-axis. By interchanging the y's and the x's we have the formula $x^2 = y$ or $y = x^2$. This too represents a parabola but one with its axis along the y-axis because the effect of the x and y interchange is merely to turn the curve through $90°$. The two forms of the curve are shown in Fig. 67.

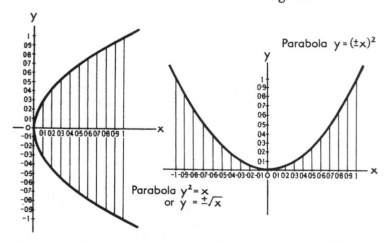

FIG. 67.

In Archimedes' method a right-angled triangle was inscribed in the segment, or rather in the half-segment. In Fig. 68 is shown this triangle drawn in the parabola, now in its new position.

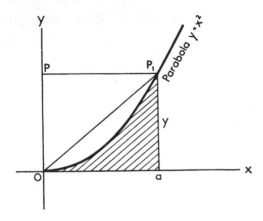

FIG. 68

For the integration the limits will be 0 and a. For the shaded part of Fig. 68 then, the curve being $y = x^2$, we have

$$\text{Area of segment} = \int_0^a x^2 dx = \left[\frac{x^3}{3}\right]_0^a = \frac{a^3}{3} - \frac{0}{3}$$

$$= \frac{a^3}{3}.$$

To find the area that Archimedes calculated we must subtract the shaded area from the rectangle which has base a and height y. When $x = a$, $y = x^2 = a^2$, so that the area of the rectangle is $a \cdot y = a \cdot a^2 = a^3$. Therefore the area of the convex parabolic segment is $a^3 - \frac{1}{3} \cdot a^3 = \frac{2}{3}a^3$. So the area of the segment is two-thirds of the area of the rectangle in which it stands—in exact agreement with Archimedes' result ; we honour the name of the pioneer of the calculus!

Although we went to the trouble of interchanging the x and the y in the equation of the parabola it is quite possible to use the integration method without making this change. The only advantage of the change is that irrational numbers are avoided. So we will repeat the quadrature of the parabola once more, this time using the equation in its first form, $y^2 = x$ or $y = \pm\sqrt{x}$. The $+$ and $-$ signs show that this parabola has one branch above and one below the x-axis. As we have done hitherto, we shall deal only with one-half of the parabolic segment and will take the positive sign, *i.e.*, that branch of the curve in the first quadrant. We will let the limits be 0 and b. Then

$$\text{Area} = \int y \,.\, dx = \int_0^b \sqrt{x}\,dx$$

$$= \int_0^b x^{\frac{1}{2}}dx = \tfrac{1}{\frac{3}{2}}\big[x^{\frac{1}{2}+1}\big]_0^b$$

$$= \tfrac{2}{3}(b^{\frac{3}{2}} - 0^{\frac{3}{2}})$$

$$= \tfrac{2}{3}b^{\frac{3}{2}} \text{ or } \tfrac{2}{3}b\sqrt{b}.$$

When $x = b$, $y = \sqrt{b}$, so that the area of the circumscribing rectangle is $b \,.\, \sqrt{b}$. In other words, we have shown again that the area of the parabolic segment is exactly $\tfrac{2}{3}$ of the area of the rectangle in which it fits. Again we must acknowledge the genius of Archimedes, who first proved this fact.

Now we are sufficiently practised in the methods of the calculus to attempt the integration of a negative power of x, namely x^{-1} or $\dfrac{1}{x}$. The equation $y = \dfrac{1}{x}$ is that of a hyperbola ; this can be seen if we substitute values of x, find the corresponding ordinates and plot the points as a graph. There are some interesting things about it. When x is large, y is small ; when $x = 1$, $y = 1$ also ; when x is smaller than 1, y is greater than 1. However large x is made y is always finite ; although it gets smaller and smaller it never disappears completely. However small x may become, y is always finite but gets larger and larger. The two axes are said to be " asymptotes " of the curve, they are the " tangents at infinity."

Let us first make some remarks in passing about the origin of the hyperbola. It is one of the important class of curves (circle, ellipse, parabola, hyperbola), which are represented by equations of the second degree in x and y. Our equation $y = \dfrac{1}{x}$ is a quadratic function though it may not look like it at first sight. Multiplying it by x, we have $xy = 1$ and xy is a term of one degree in x and one degree in y, therefore of the second degree as a whole. These curves of the second degree are called " Conic Sections " in geometry.

In Fig. 69 a cone is cut in four different ways. The resulting " sections " of the cone are the curves we have just mentioned.

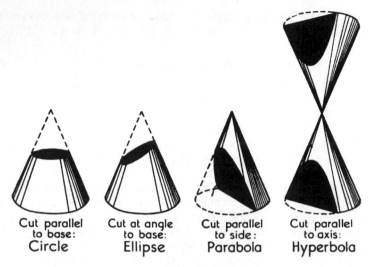

Cut parallel to base: **Circle**

Cut at angle to base: **Ellipse**

Cut parallel to side: **Parabola**

Cut parallel to axis: **Hyperbola**

Fig. 69.

The circle and ellipse are closed curves, the parabola is a curve of one branch and the hyperbola of two branches, each branch opening out wider and wider.

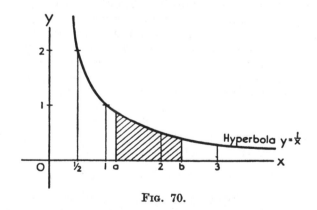

Fig. 70.

In Fig. 70 part of the curve $y = \dfrac{1}{x}$ is drawn ; we are trying to find the area under the curve which lies between the ordinates at a and b. It looks straightforward enough ; we will therefore apply the formula

$$\text{Area} = \int_a^b y\,dx = \int_a^b \frac{1}{x}\,dx = \int_a^b x^{-1}\,dx.$$

We know that $\int x^m dx = \dfrac{x^{m+1}}{m+1}$, but on applying this formula now, we get $\dfrac{x^{-1+1}}{-1+1} = \dfrac{x^0}{0} = ?$. This does not look promising. The reason is that the formula we have used, $\int x^m dx = \dfrac{x^{m+1}}{m+1}$, is valid for every value of x except the one in which we are interested at the moment, namely $m = -1$. The formula should therefore be written

$$\int x^m dx = \frac{x^{m+1}}{m+1}\,[m \neq -1],$$

the sign $\neq$ standing for " not equal to."

We have deliberately led the reader on to this thin ice to give him his first shock over the capabilities of the integral calculus. The answer is known, however, but for many the logical difficulties of the problem led to criticism of the calculus and to doubts about its validity. The secret lies in the word " logarithm." Though we have achieved our aim of leading the reader from the beginnings of arithmetic to the calculus, we must not leave unsolved the problem of integrating x^{-1}.

A last word about rectification ; we have shown that the length of a curve can be represented by the integral $\int \sqrt{(dx)^2 + (dy)^2}$. We have given no numerical examples of this because the rectification integrals are all too difficult for our present knowledge. To show the reader what can happen we will just give the rectification formula for the simple case of the parabola $y = kx^2$. The length of the curve from $x = 0$ to any point whose abscissa is x, is equal to $\int_0^x \sqrt{1 + 4k^2x^2}\,.\,dx$. The result of this integration is the formidable function

$$S = \tfrac{1}{2}x\sqrt{1 + 4k^2x^2} + \frac{1}{4k}\log_e(\sqrt{1 + 4k^2x^2} + 2kx),$$

where k can be a number of any kind, depending on the actual parabola considered.

CHAPTER XXXVI

LOGARITHMS

THE word " logarithm " comes from " logos arithmos," Greek for " ratio number." Logarithms were introduced by Sir John Napier in 1614 after more than twenty years' work on their theory and in computing tables of them. We will not spend any more time on the history of the subject, however, but will immediately attack this, our last problem, by direct assault.

We have already studied functions of the kind x^n in which n has taken various constant values, e.g., x^0, x^1, x^2, x^3 . . . x^n ; these are " powers " of x. We have also studied the inverse functions, that is, roots of x, which we can represent as $y^n = x$, or $y = \sqrt[n]{x}$. It is clear that in these functions the exponent n has taken a selection of values, mostly whole numbers and unit fractions, so that it is not unreasonable now if we consider the function $y = a^x$, in which the exponent, x, has become variable. This function $y = a^x$ is called the " exponential function." We will illustrate it for the case in which $a = 5$. When $x = 1$; $y = 5^1$; $x = 2$, $y = 5^2 = 25$; $x = 3$, $y = 5^3 = 125$; $x = 4$, $y = 5^4 = 625$, and so on. From the exponential function $y = a^4$ arises the inverse question, " To what power must the constant a, here equal to 5, be raised to give the result 125 ? " The answer is obviously given by $\sqrt[x]{125} = 5$, but we have no general means of finding the xth root of a number. Though the reader may be able to solve the last equation, the hopelessness of our present situation will be brought home to him if he tries to solve $10^x = 2$ in the same way.

The exponential function then is written in the form $y = a^x$, where a is a constant and x is a known variable. The inverse question arises, " given a and y in this equation, what is x ? " The answer to this question is written as

$$x = \log_a y$$

and we read it as " x is the logarithm of y to the base a." To what power must 10 be raised to give the result 2 ?

If $\qquad\qquad 10^x = 2,$

x is the " logarithm " of 2 to the base 10, or

$$x = \log_{10} 2 = \cdot 30103 \ldots$$

as we can find from the tables.   Hence

$$10^{\cdot 30103\cdots} = 2.$$

At this point we might very well review the various operations which we have discussed in this book and construct a table which shows them as falling into two classes ; they either " build up " numbers, i.e., they are " synthesising " operations, or they " break down " numbers, i.e., they are " analysing " operations.   Each synthesising operation has its analysing counterpart :—

Synthesising Operation	*Analysing Operation*
1. Addition—$(a + b)$	Subtraction—$(a - b)$.
2. Multiplication—$(a \cdot b)$	Division—$(a \div b)$.
3. Raising to a power—(x^n)	Extraction of roots—$(\sqrt[n]{x})$.
4. Raising to the exponential—(a^x)	Taking the logarithm—$(\log_a x)$.
5. Integration—$\int f'(x)\, dx$	Differentiation—$\left(\dfrac{dy}{dx} = f'(x)\right)$.

With this table complete we can say that we have at least a bowing acquaintance with the operations listed ; we should know what the commands mean, but in the case of the last operation to be considered, taking the logarithm, we still do not understand.

There are two points about logarithms that we must now consider briefly.   Firstly, what kind of numbers are they ? and secondly, how are they used ?   Logarithms are in general irrational numbers.   We do not propose to show in this book how logarithms are calculated, though these calculations are tedious and heavy rather than difficult.   For all practical requirements books of " log. tables " are available which provide the logarithms of numbers correct to 4, 5 . . . 7 . . . decimal places.   The use of such tables for finding powers, roots and simplifying multiplication and division is fully described in any good arithmetic book ; we are here more concerned with the underlying theory of logarithms.

As the base, a, of a system of logarithms any number may be used.   In practice, however, only two base numbers are of importance. The number 10 is the base of Briggs's logarithms and the number e is the base of the " natural " or Napierian logarithms. There is rarely any doubt about which of these two systems is being used, even though the logarithm, $e.g.$, $\log x$, may not always show whether e or 10 is the base. In practical calculations the base 10 is mostly used ; in theoretical work, the base e. In this chapter we shall always indicate which base is being used, $e.g.$, $\log_{10} x$, $\log_e x$; or $\log_a x$ to indicate that the base is any constant number " a." Sometimes $\log_e x$ will be found written as $ln\ x$. (ln is short for " logarithmus naturalis.")

If we have an equation $c = d$, it remains an equation if we take logarithms of both sides, so that $\log_a c = \log_a d$. Similarly, if we begin with $\log_a c = \log_a d$, it follows then that $c = d$.

The fundamental property of logarithms is expressed in the equation
$$\log_a (c \cdot d) = \log_a c + \log_a d,$$
which is, expressed in words, that the logarithm of the product of two numbers is equal to the sum of the logarithms of the two numbers. Let us deduce this property. Suppose $\log_a c = C$ and $\log_a d = D$, then we can write these in the exponential form
$$c = a^C \text{ and } d = a^D.$$

The product $c \cdot d$ is therefore equal to the product $a^C \cdot a^D = a^{C+D}$. Therefore we can write
$$cd = a^{C+D}.$$

This $(C + D)$ is the exponent, to the base a, of the product $c \cdot d$; it is therefore the logarithm of $c \cdot d$ to the base a. Hence
$$\log_a cd = C + D$$
but $C = \log_a c$ and $D = \log_a d$ by our definition, so that finally
$$\log_a cd = \log_a c + \log_a d.$$

Using a similar method of working out the equation in exponentials and indices and then taking logarithms, the reader can verify the following additional rules for logarithms :—

For division :—
$$\log_a \frac{c}{d} = \log_a c - \log_a d.$$

For raising to a power :—
$$\log_a c^d = d \log_a c.$$

For taking a root :—

$$\log_a \sqrt[d]{c} = \frac{1}{d} \log_a c.$$

With this brief outline of the main properties of logarithms we must go on to study more closely the particular properties of natural logarithms for they have a crucial importance in the development of the differential and integral calculus. The rules we have just summarised hold for the base e as for any other number, so that

$$\log_e ab = \log_e a + \log_e b,$$

and so on.

Rather abruptly we are going to investigate a new function, namely $\left(1 + \dfrac{1}{n}\right)^n$ where n is a positive whole number. Let us begin by giving n simple values which make the function easy to calculate. We will tabulate the results correct to four decimal places :—

n	$\left(1 + \dfrac{1}{n}\right)^n$	Result
1	$(1 + 1)^1$	2·0000
2	$(1 + \frac{1}{2})^2 = (1\frac{1}{2})^2 =$	2·2500
3	$(1 + \frac{1}{3})^3 = (1\frac{1}{3})^3 =$	2·3737
4	$(1 + \frac{1}{4})^4 = (1\frac{1}{4})^4 =$	2·4383
5	$(1 + \frac{1}{5})^5 = (1\frac{1}{5}^5) =$	2·4883

As n increases from 1 to 5 the value of the function increases but with smaller and smaller increments. The question arises whether $\left(1 + \dfrac{1}{n}\right)^n$ continues to increase indefinitely as n increases or whether it " tends to a limit." Let us make a big jump in the value of n ; suppose we put $n = 100$. If we work out the value of $(1 + \frac{1}{100})^{100} = 1{\cdot}01^{100}$ by logarithms (7-figure) or by the Binomial Theorem, we find that the result is 2·7048. . . . This strengthens the suspicion that the function is tending to a limit rather like the infinite geometric series we have considered, e.g., the series $1 + \frac{1}{2} + \frac{1}{4} + \frac{1}{8} \ldots$ we found to tend towards 2·0000 as the number of terms taken were increased ; there is no possibility of the series exceeding 2·0000 . . . however many terms are taken. If we make another jump in the value of n we find that for $n = 1,000,000$ the function $(1{\cdot}000001)^{1,000,000}$ is equal to 2·718. . . . It is clear that the function deserves a closer analytical inspection.

Let us take the general case $\left(1 + \dfrac{1}{n}\right)^n$ and expand it as a finite series using the Binomial Theorem. Then

$$\left(1 + \frac{1}{n}\right)^n = 1 + {}_nC_1 \cdot \frac{1}{n} + {}_nC_2 \cdot \frac{1}{n^2} + {}_nC_3 \cdot \frac{1}{n^3} + \ldots \frac{1}{n^n}$$

$$= 1 + n \cdot \frac{1}{n} + \frac{n(n-1)}{1 \cdot 2} \cdot \frac{1}{n^2} + \frac{n(n-1)(n-2)}{1 \cdot 2 \cdot 3} \cdot \frac{1}{n^3} + \ldots \frac{1}{n^n}$$

$$= 1 + \frac{1}{1} + 1\left(1 - \frac{1}{n}\right) \cdot \frac{1}{2!} + 1\left(1 - \frac{1}{n}\right)\left(1 - \frac{2}{n}\right) \cdot \frac{1}{3!} + \ldots \frac{1}{n^n}.$$

Now when n is very large we can say that $\dfrac{1}{n}$ is negligibly small, or in symbols, an $n \to \infty$, $\dfrac{1}{n} \to 0$, so that we can write the *first few terms* of the series as $1 + \dfrac{1}{1!} + \dfrac{1}{2!} + \dfrac{1}{3!} + \ldots$ as all the terms of the form $\left(1 - \dfrac{r}{n}\right) \to 1$. It does *not* follow from this analysis that we have *proved* that the *whole* series can be written as

$$\underset{n \to \infty}{L}t\left(1 + \frac{1}{n}\right)^n = 1 + \frac{1}{1!} + \frac{1}{2!} + \frac{1}{3!} + \ldots \frac{1}{n!} + \ldots$$

For example, the last term of the binomial expansion is ${}_nC_n \cdot \dfrac{1}{n^n} = 1 \cdot \dfrac{1}{n^n}$, which is clearly not the same thing as $\dfrac{1}{n!}$ as our series requires. However, the reader will agree that there is a strong presumption in favour of the series as it has just been written. For a complete and rigorous proof he must consult a specialist text-book of analysis, for example, Hardy's *Pure Mathematics*.

We must now interest ourselves in the sum of the series. Multiplying out the factorial terms it becomes

$$1 + 1 + \tfrac{1}{2} + \tfrac{1}{6} + \tfrac{1}{24} + \tfrac{1}{120} \ldots$$

It is clear that the sum is greater than 2·5, which is merely the sum of the first three terms. Can we find an upper limit for it ? Yes, if we ignore for a moment the first term and compare the series

$$1 + \tfrac{1}{2} + \tfrac{1}{6} + \tfrac{1}{24} + \tfrac{1}{120} + \ldots$$

with the series

$$1 + \tfrac{1}{2} + \tfrac{1}{4} + \tfrac{1}{8} + \tfrac{1}{16} + \ldots$$

which we already know as a geometric series for which $S_\infty = \dfrac{1}{1 - \frac{1}{2}} = 2$. In the series we are investigating, every term is less than the corresponding term of the geometric series. Hence we know that

$$1 + \tfrac{1}{2} + \tfrac{1}{6} + \tfrac{1}{24} + \tfrac{1}{120} + \ldots \text{ to infinity} < 2,$$

and therefore that

$$1 + \frac{1}{1!} + \frac{1}{2!} + \frac{1}{3!} + \frac{1}{4!} \ldots$$

lies between 2·5 and 3·0. In other words, the value of $\left(1 + \dfrac{1}{n}\right)^n$ tends to a limit as $n \rightarrow \infty$. This limit is also the limiting sum of the infinite series $1 + \dfrac{1}{1!} + \dfrac{1}{2!} + \dfrac{1}{3!} + \dfrac{1}{4!} \ldots$ which we can calculate term by term to any degree of accuracy we may wish. The numerical value of this sum is 2·71828182845904...

This number, like π, is irrational and is represented on Euler's suggestion by the letter "e." It is one of the key numbers of mathematics. We must hasten on and discover why it is so important.

Having expressed e as an infinite series, we are going to show that its exponential, e^x, can also be expressed as an infinite series. We have then

$$e = \mathop{Lt}_{n \to \infty}\left(1 + \frac{1}{n}\right)^n,$$

so that we can write

$$e^x = \left[\mathop{Lt}_{n \to \infty}\left(1 + \frac{1}{n}\right)^n\right]^x$$

$$= \mathop{Lt}_{n \to \infty}\left(1 + \frac{1}{n}\right)^{nx}.$$

Once again we expand the binomial and get

$$\left(1 + \frac{1}{n}\right)^{nx} = 1 + nx \cdot \frac{1}{n} + \frac{nx \cdot (nx - 1)}{1 \cdot 2} \cdot \frac{1}{n^2} +$$

$$\frac{nx(nx - 1)(nx - 2)}{1 \cdot 2 \cdot 3} \cdot \frac{1}{n^3} + \ldots \frac{1}{n^{nx}}.$$

As long as x is finite we can say as before that

$$\frac{nx(nx-1)}{1 \cdot 2}\frac{1}{n^2} = x\left(x - \frac{1}{n}\right) \cdot \frac{1}{2!}$$

$$\rightarrow \frac{x^2}{2!} \text{ as } n \rightarrow \infty$$

and so on for subsequent terms.

We therefore have, with the same limitations in our " proof " as we had before,

$$e^x = 1 + \frac{x}{1!} + \frac{x^2}{2!} + \frac{x^3}{3!} + \frac{x^4}{4!} + \cdots$$

This series can also be shown to have a " limiting value " for any finite value of x. It is called the " exponential series." It is a polynomial function of x, so let us differentiate it term by term :—

$$y = e^x = 1 + \frac{x}{1!} + \frac{x^2}{2!} + \frac{x^3}{3!} \cdots$$

$$\frac{dy}{dx} = \frac{d(e^x)}{dx} = 0 + \frac{1}{1!} + \frac{2x}{2!} + \frac{3x^2}{3!} + \cdots$$

$$= \frac{1}{1} + \frac{2x}{1 \cdot 2} + \frac{3x^2}{1 \cdot 2 \cdot 3} + \frac{4x^3}{1 \cdot 2 \cdot 3 \cdot 4} + \cdots$$

$$= 1 + \frac{x}{1!} + \frac{x^2}{2!} + \frac{2x^3}{3!} + \cdots$$

which is the series we began with! In other words, the differential coefficient of e^x is e^x. Perhaps we may now begin to see why e is so important a number in higher mathematics. By our integration formula $\int y' dx = y$, therefore $\int e^x \cdot dx = e^x + C$. The results make e^x a very simple function to deal with.

Having established these properties of the number e and of its exponential function, we must now return to the logarithmic function, this time with e as base,

that is, $y = \log_e x.$

If y is the logarithm of x to the base e then we have

$$x = e^y,$$

an equation in which we can consider x to be the dependent and y the independent variable. Now let us differentiate this equation " with respect to y," that is, find $\dfrac{dx}{dy}$ instead of the more usual $\dfrac{dy}{dx}$:—

$$x = e^y$$

$$\frac{dx}{dy} = e^y.$$

As $e^y = x$, it follows that $\dfrac{dx}{dy} = x$. From this result we can obtain $\dfrac{dy}{dx}$, since $\dfrac{dy}{dx}$ and $\dfrac{dx}{dy}$ are simply reciprocal, that is $\dfrac{dy}{dx} = \dfrac{1}{\dfrac{dx}{dy}}$ and as $\dfrac{dx}{dy} = x$, $\dfrac{dy}{dx} = \dfrac{1}{x}$.

By this roundabout method we have established the surprising result that the differential coefficient of $\log_e x$ is $\dfrac{1}{x}$, and $\dfrac{1}{x} = x^{-1}$ in the index notation. Notice that when we differentiate x^3 we get $3x^2$, from x^2 we get $2x^1$, from x we get 1 or x^0, from 1 or x^0 we get 0, from x^{-1} we get $-1 \cdot x^{-1-1}$ or $-x^{-2}$. There is no power of x which gives the differential coefficient x^{-1}. Herein lies the solution of the difficulty in the integration rule that $\int x^m dx = \dfrac{1}{m+1} x^{m+1}$ except for the value $m = -1$, which was the case we met in attempting the quadrature of the hyperbola. As the function which on differentiation gives x^{-1} is $\log_e x$, we have found that for the exceptional case $m = -1$,

$$\int x^m dx = \int x^{-1} dx = \log_e x + C.$$

Therefore the determinate integral which expressed the area under the hyperbola can now be completed :—

$$\text{Area} = \int_a^b x^{-1} dx = \left[\log_e x\right]_a^b = \log_e b - \log_e a.$$

By the second of the logarithm rules this result can be expressed differently because

$$\log_e b - \log_e a = \log_e \frac{b}{a}.$$

If therefore we integrate between the limits $x = 2a$, $x = a$, we should have the area $= \log_e 2a - \log_e a = \log_e \dfrac{2a}{a} = \log_e 2$; if the limits were $2b$ and b, the area would again be $\log_e 2$.

It is often convenient to express the constant of the indefinite integral as a logarithm itself. If C is a constant it is the logarithm of another constant number, let us say c. Hence

$$\int x^{-1} dx = \log_e x + C$$
$$= \log_e x + \log_e c$$
$$= \log_e cx,$$

and this is another way of writing the general integral.

We will just add that one system of logarithms can of course be converted to another. In pure mathematics logarithms to the base e are used for reasons which should now be obvious. Most practical work, as we have said before, is done with logarithms based on 10, but occasionally we need to change from one system to another. To convert natural logarithms to those based on 10 we use the formula

$$\log_{10} x = 0{\cdot}4343 \log_e x,$$

and reciprocally, $\log_e x = 2{\cdot}303 \log_{10} x.$

In conclusion we will give a picture of the curve $y = \log_e x$, the logarithmic curve (Fig. 71). The logarithm of 1 is 0 whatever the base of the system may be, so that when $x = 1$, $y = 0$, and therefore the curve cuts the x-axis at $x = 1$. When $x = e$, $y = \log_e e = \log_e e^1 = 1$;   when $x = e^2$, $y = \log_e e^2 = 2$; when $x = e^3$, $y = 3$, and so on. Similarly, when $x = e^{-\frac{1}{2}}$, $y = -\frac{1}{2}$; $x = e^{-\frac{1}{3}}$, $y = -\frac{1}{3}$, and so on. The graph rises steeply when x lies between 0 and 1, and thereafter rises more and more slowly, increasing always as x increases with no asymptote like the hyperbola.

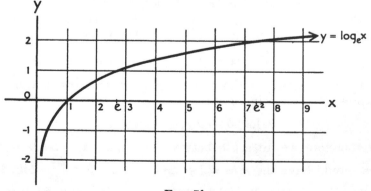

FIG. 71.

CHAPTER XXXVII

INTERPOLATION, EXTRAPOLATION, CONCLUSION

On looking up logarithms in the tables, we may sometimes find that the value we want lies somewhere between two values given in the table—obviously our tables cannot give us the logarithm of every number in the number line. This search for intermediate values is called " interpolation," from the Latin " interpolare," to put in. As this idea is of great importance in all branches of mathematics and science in which tabulated values are used, we must discuss it briefly. The subject in fact is too big and too difficult for us to do otherwise, but we can answer this question, " What is an interpolated value ? " both by means of arithmetic and the geometry we already know. Arithmetically, an interpolated value is a number which lies between two numbers already known, but it lies at a particular point in the gap between them. Suppose we know that the logarithm of 1499 is 3·17580 and that the logarithm of 1500 is 3·17609 ; we might need the logarithm of 1499·5, which is not given by our tables. Now 1499·5 is mid-way between 1499 and 1500. Commonsense would lead us to expect that log 1499·5 is therefore mid-way between 3·17580 and 3·17609, that is, it is 3·17595 correct to five decimal places. In a similar way we could " interpolate " for another value. Thus, to find the logarithm of 1499·8 we must divide the " difference " between the two logarithm values into ten parts and then add eight of these tenths to the smaller logarithm. The difference is 3·17608 — 3·17580, or ·00029, and one-tenth of this is ·000029. Eight such tenths make ·000232. So the logarithm of 1499·8 is equal to 3·17580 + ·000232, or 3·17603 correct to five decimal places.

Interpolation in logarithm tables, generally speaking, follows the method we have just demonstrated. It is achieved by the use of " proportional parts "—the same thing as our tenths. There is however an important condition which has to be fulfilled for the correct application of this method, namely that the values in the gap between two known numbers increase steadily and uniformly ; this is the condition required for the so-called " linear " interpolation. To make the idea of interpolation clearer let us now examine it from the geometrical point of view.   Treating this analytically we mark off numbers along the x-axis and draw at given values of x ordinates equal to the corresponding function values of y given by the table, in this case the logarithms.   Of course, if we knew the equation of the function not only could we draw the graph completely but our interpolation would no longer be needed.   To get an intermediate value all we need to do is to

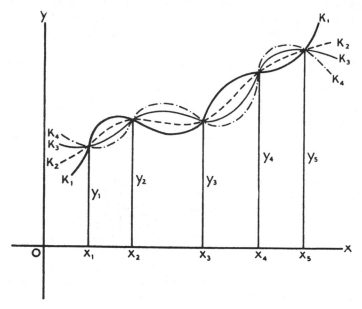

FIG. 72.

substitute the x in the "$f(x)$," whatever that may be, and work out the result.   However, we must assume here that we know the values of $f(x)$ only for some particular values of x. Our "graph" therefore consists of a few isolated points and we do not know what happens in between them.   We now assume that the values we need lie on a curve which is continuous across the gap from one isolated point to the next. On looking at Fig. 72 it can be seen, however, that our troubles are not at an end because we can draw as many curves as we like through any number of isolated points.

The problem of interpolation is therefore insoluble unless

we make certain assumptions which we will now discuss. Suppose we have to interpolate between only two points of the curve which is yet unknown.

If we choose the "linear" interpolation we simply draw a straight line between the two points. We now imagine that the gaps between the ordinates are divided into a large number of small parts, each of equal width Δx. The total gap will then be $\Sigma \Delta x$—the sum of all these parts. Corresponding to each Δx there will be a Δy, so that $\Sigma \Delta y$ is the difference in the heights of the two ordinates ; and for the straight line we are considering all the Δy's will also be equal. We shall therefore have this proportion :—

$$\frac{\Sigma \Delta x}{\Sigma \Delta y} = \frac{\Delta x}{\Delta y}.$$

Hence, if there are n parts in the total sum

$$\frac{1}{n}\Sigma \Delta x = \Delta x$$

and

$$\frac{1}{n}\Sigma \Delta y = \Delta y.$$

This kind of proportional interpolation is permissible only if we can assume that the "curve" is a straight line. This method is used in finding logarithms though we have already seen in Fig. 71 that the logarithmic graph is a curve : but for large numbers the curve is so nearly a straight line that the error is negligible.

There are of course other methods of interpolation if it cannot be assumed that the curve is linear in the gap or "interval" with which we are concerned. It is impossible to pursue these methods here, and for elementary work they are not really necessary. In most parts of most mathematical tables linear interpolation is accurate enough for general use.

But "grey is all theory" and it is time we gave a practical example to show when interpolation is required. Let us suppose that in a certain country a census of population is taken every ten years, say in 1910, 1920, 1930, and so on. From the results of the census it is possible to estimate by interpolation methods not only what the population was in 1927 but also what it was in 1895 or what it will be in 1960. This extension of the field beyond the known range is called

" extrapolation," but its methods of calculation (when it is possible) are basically the same as those of interpolation.

It is by extrapolation that modern astronomers can predict eclipses of the sun and moon, and determine very accurately the dates of eclipses mentioned in ancient historical documents. Indeed it should be added that our mathematical ideas of interpolation and extrapolation are merely a formal way of expressing an idea without which we should find life very difficult. We assume that the pattern of our lives, of business,

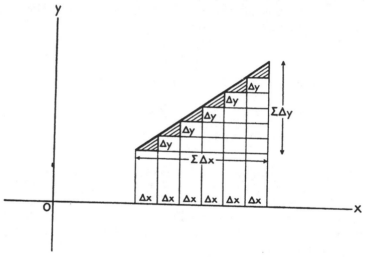

Fig. 73.

of politics, will continue in the future much the same as they have proceeded in the past. It is a pity that we cannot learn from the cautious mathematician what risks there are in extrapolating too firmly.

By now we have more than completed the task which we set ourselves at the beginning of the book. We hope we have kept the promise we gave then, to lead the reader from the multiplication table to the integral calculus. At times, especially in the later chapters, there has been a marked tendency to say that this or that problem was beyond the scope of this book. Many readers must have felt annoyed and frustrated. Yet these references to the limitation of the content of the book were intended. We need never be depressed by

the realisation of our limitations. Consciousness of the lack
of knowledge is surely the greatest spur to acquiring more
knowledge. Hard work will help us on. The extent of a man's
future can be measured by his consciousness of unsolved
problems and by the way in which these problems constantly
exercise his mind. This is true of the future of a nation and
indeed of mankind itself. If we think we have progressed
enough and have fulfilled our purpose, we are doomed to
atrophy. Each one of our mathematical heroes, whether he
was called Pythagoras, Eudoxus, Euclid, Archimedes,
Apollonius, Alchwarizmi, Kepler or Descartes, imagined that
he had completed the development of his subject and his
contemporaries agreed with their intellectual leader. But
then came Newton, Leibniz, Euler, Lagrange, Gauss, Riemann,
Weierstrass, Minkowski, Hilbert. And there are more great
men to come ; there always will be—in mathematics too.
So that if we have come to know only a small part of the realm
of mathematics, we can yet say with pride that we have
thoroughly investigated something that is most essential. We
do at least realise the range of mathematics and what there is
to be found in it, to be learnt and to be analysed. We have
gained something that is most important (more important
than a knowledge of mathematics), namely respect for the
greatness of the human mind. There is a quotation from
Faust which runs : " We are accustomed to find that men
despise what they do not understand." If we may treat
Goethe's words as an " inverse function," we can say that
men only respect what they understand. Respect for an object
implies that we value it. All must strive after what is
valuable, not in our own selfish interests, but to the greater
glory of the human spirit.

ERRATA

PAGE, LINE	REPLACE	WITH
199, 3	$+\dfrac{(-2)(-3)(-4)}{1.2.3}(-x)^4+$	$+\dfrac{(-2)(-3)(-4)}{1.2.3}(-x)^3+$
205, 22	$P_2N_2=\dfrac{1}{2}P_1N_2$	$P_2N_2=\dfrac{1}{2}P_1N_1$
226, 2	$ff'(x)\,dx=f(x)+c$	$\int f'(x)\,dx=f(x)+c$
246, 8	an $n\to\infty$	as $n\to\infty$
248, 15	$=+\dfrac{x}{1!}+\dfrac{x^2}{2!}+\dfrac{2x^3}{3!}+$	$=+\dfrac{x}{1!}+\dfrac{x^2}{2!}+\dfrac{x^3}{3!}+$

A CATALOG OF SELECTED
DOVER BOOKS
IN ALL FIELDS OF INTEREST

A CATALOG OF SELECTED DOVER
BOOKS IN ALL FIELDS OF INTEREST

CONCERNING THE SPIRITUAL IN ART, Wassily Kandinsky. Pioneering work by father of abstract art. Thoughts on color theory, nature of art. Analysis of earlier masters. 12 illustrations. 80pp. of text. 5⅜ x 8½. 23411-8

ANIMALS: 1,419 Copyright-Free Illustrations of Mammals, Birds, Fish, Insects, etc., Jim Harter (ed.). Clear wood engravings present, in extremely lifelike poses, over 1,000 species of animals. One of the most extensive pictorial sourcebooks of its kind. Captions. Index. 284pp. 9 x 12. 23766-4

CELTIC ART: The Methods of Construction, George Bain. Simple geometric techniques for making Celtic interlacements, spirals, Kells-type initials, animals, humans, etc. Over 500 illustrations. 160pp. 9 x 12. (Available in U.S. only.) 22923-8

AN ATLAS OF ANATOMY FOR ARTISTS, Fritz Schider. Most thorough reference work on art anatomy in the world. Hundreds of illustrations, including selections from works by Vesalius, Leonardo, Goya, Ingres, Michelangelo, others. 593 illustrations. 192pp. 7⅛ x 10¼. 20241-0

CELTIC HAND STROKE-BY-STROKE (Irish Half-Uncial from "The Book of Kells"): An Arthur Baker Calligraphy Manual, Arthur Baker. Complete guide to creating each letter of the alphabet in distinctive Celtic manner. Covers hand position, strokes, pens, inks, paper, more. Illustrated. 48pp. 8¼ x 11. 24336-2

EASY ORIGAMI, John Montroll. Charming collection of 32 projects (hat, cup, pelican, piano, swan, many more) specially designed for the novice origami hobbyist. Clearly illustrated easy-to-follow instructions insure that even beginning papercrafters will achieve successful results. 48pp. 8¼ x 11. 27298-2

THE COMPLETE BOOK OF BIRDHOUSE CONSTRUCTION FOR WOODWORKERS, Scott D. Campbell. Detailed instructions, illustrations, tables. Also data on bird habitat and instinct patterns. Bibliography. 3 tables. 63 illustrations in 15 figures. 48pp. 5¼ x 8½. 24407-5

BLOOMINGDALE'S ILLUSTRATED 1886 CATALOG: Fashions, Dry Goods and Housewares, Bloomingdale Brothers. Famed merchants' extremely rare catalog depicting about 1,700 products: clothing, housewares, firearms, dry goods, jewelry, more. Invaluable for dating, identifying vintage items. Also, copyright-free graphics for artists, designers. Co-published with Henry Ford Museum & Greenfield Village. 160pp. 8¼ x 11. 25780-0

HISTORIC COSTUME IN PICTURES, Braun & Schneider. Over 1,450 costumed figures in clearly detailed engravings—from dawn of civilization to end of 19th century. Captions. Many folk costumes. 256pp. 8⅜ x 11¾. 23150-X

THE STORY OF THE TITANIC AS TOLD BY ITS SURVIVORS, Jack Winocour (ed.). What it was really like. Panic, despair, shocking inefficiency, and a little heroism. More thrilling than any fictional account. 26 illustrations. 320pp. 5⅜ x 8½.
20610-6

FAIRY AND FOLK TALES OF THE IRISH PEASANTRY, William Butler Yeats (ed.). Treasury of 64 tales from the twilight world of Celtic myth and legend: "The Soul Cages," "The Kildare Pooka," "King O'Toole and his Goose," many more. Introduction and Notes by W. B. Yeats. 352pp. 5⅜ x 8½.
26941-8

BUDDHIST MAHAYANA TEXTS, E. B. Cowell and others (eds.). Superb, accurate translations of basic documents in Mahayana Buddhism, highly important in history of religions. The Buddha-karita of Asvaghosha, Larger Sukhavativyuha, more. 448pp. 5⅜ x 8½.
25552-2

ONE TWO THREE . . . INFINITY: Facts and Speculations of Science, George Gamow. Great physicist's fascinating, readable overview of contemporary science: number theory, relativity, fourth dimension, entropy, genes, atomic structure, much more. 128 illustrations. Index. 352pp. 5⅜ x 8½.
25664-2

EXPERIMENTATION AND MEASUREMENT, W. J. Youden. Introductory manual explains laws of measurement in simple terms and offers tips for achieving accuracy and minimizing errors. Mathematics of measurement, use of instruments, experimenting with machines. 1994 edition. Foreword. Preface. Introduction. Epilogue. Selected Readings. Glossary. Index. Tables and figures. 128pp. 5⅜ x 8½.
40451-X

DALÍ ON MODERN ART: The Cuckolds of Antiquated Modern Art, Salvador Dalí. Influential painter skewers modern art and its practitioners. Outrageous evaluations of Picasso, Cézanne, Turner, more. 15 renderings of paintings discussed. 44 calligraphic decorations by Dalí. 96pp. 5⅜ x 8½. (Available in U.S. only.)
29220-7

ANTIQUE PLAYING CARDS: A Pictorial History, Henry René D'Allemagne. Over 900 elaborate, decorative images from rare playing cards (14th–20th centuries): Bacchus, death, dancing dogs, hunting scenes, royal coats of arms, players cheating, much more. 96pp. 9¼ x 12¼.
29265-7

MAKING FURNITURE MASTERPIECES: 30 Projects with Measured Drawings, Franklin H. Gottshall. Step-by-step instructions, illustrations for constructing handsome, useful pieces, among them a Sheraton desk, Chippendale chair, Spanish desk, Queen Anne table and a William and Mary dressing mirror. 224pp. 8⅛ x 11¼.
29338-6

THE FOSSIL BOOK: A Record of Prehistoric Life, Patricia V. Rich et al. Profusely illustrated definitive guide covers everything from single-celled organisms and dinosaurs to birds and mammals and the interplay between climate and man. Over 1,500 illustrations. 760pp. 7½ x 10⅛.
29371-8